Advances in Information Security

Volume 92

The purpose of the *Advances in Information Security* book series is to establish the state of the art and set the course for future research in information security. The scope of this series includes not only all aspects of computer, network security, and cryptography, but related areas, such as fault tolerance and software assurance. The series serves as a central source of reference for information security research and developments. The series aims to publish thorough and cohesive overviews on specific topics in Information Security, as well as works that are larger in scope than survey articles and that will contain more detailed background information. The series also provides a single point of coverage of advanced and timely topics and a forum for topics that may not have reached a level of maturity to warrant a comprehensive textbook.

Anne V. D. M. Kayem

Enabling Privacy Preserving Data Analytics

Springer

Anne V. D. M. Kayem
Exeter, UK

ISSN 1568-2633 ISSN 2512-2193 (electronic)
Advances in Information Security
ISBN 978-3-031-93905-1 ISBN 978-3-031-93906-8 (eBook)
https://doi.org/10.1007/978-3-031-93906-8

This Springer imprint is published by the registered company Springer Nature Switzerland AG
The registered company address is: Gewerbestrasse 11, 6330 Cham, Switzerland

To my sisters . . .

Preface

The advent of legislation such as GDPR (European General Data Protection Regulation) has emphasised the need for a clear definition and understanding of digital privacy in an increasingly interconnected digital world. From the layman's and dictionary perspective, privacy can be defined very broadly as "*...the ability of an individual or group to seclude themselves or information about themselves, and thereby express themselves selectively*". In other words, privacy is a subjective concept in which one person's perception of what is private might differ considerably from another person's. The digital sphere, however, requires that we think about privacy as a collective concept in which decisions as to what should be kept private are interdependent. This raises the question of what role data dependencies play in the preservation (or violation) of the privacy of individual users.

Several approaches have been proposed to address the problem of generating privacy preserving data, each with the aim of preventing private information disclosure while at the same time maintaining a high degree of data utility. There are two main tenets to addressing the issue of generating privacy preserving data. The first is to transform the data syntactically to conceal information that might potentially enable re-identification attacks. The second is to evaluate what semantic interpretations might arise from studying the data and to distort the data to retain its salient meaning without revealing private information pertaining to individuals in the dataset. While both approaches have been widely studied and applied in practice, one of the caveats of the second approach is that use cases for the anonymised dataset must be specified apriori. In time critical and high processing scenarios, this can pose a performance bottleneck. Syntactic approaches to offer a better alternative performance wise, since the anonymised datasets are generated apriori and used on-demand. Furthermore, privacy legislation such as the European Union's GDPR (General Data Protection Regulation) place a strong emphasis on syntactic data transformation procedures to ensure user privacy. As such, while we discuss both syntactic and semantic approaches to generating privacy preserving data, our primary focus in this monograph is on syntactic approaches.

We discuss the implications of failed digital privacy on the perpetration of online harms such as fraud enabled by subverting identity management schemes. We then proceed to study the considerations that need to accounted for in discovering exposures of personally identifying information in digital settings, and the implications of transforming personal data to enforce privacy. In particular, we discuss how techniques such as outlier detection, Bayesian inference, and deep learning could potentially offer a means of analysing datasets on-the-fly to check for privacy disclosure risks and vulnerabilities.

Exeter, UK
July 2024

Anne V. D. M. Kayem

Acknowledgments I would like to thank my mentor and adviser, Prof. Dr. Christoph Meinel, for his support in conducting the research on which this monograph is based. His invaluable advice was and continues to be a strong driving force for me. Thank you for pushing me on ever so gently but very firmly.

To my family for their support and patience with me during this time—thank you. Thank you most especially for being patient with me when I was grumpy and for the "alarm-clock" throughout the writing journey, and for encouraging me on the bleak and gloomy days.

Over the years I have benefited from the insights and advice of numerous colleagues and mentors. An incomplete list in no particular order includes: Maurice Tchuente, Steve Marsh, Stephen Wolthusen, Katrin Frank, Martin Olivier, Robert Hirschfeld, Leonard Barolli, Selim Akl, Patrick Martin, Wendy Powley, Dieter Hutter, Berthold Hoffmann, Bernd Krieg-Brueckner, Felix Naumman, Andreas Polze, Cheng Feng, Haojing Yang, Antoine Bagula, Andrew Adamatsky, Djamal Djenouri, and Sylvia Osbourne.

I would also like to extend a special thank you to my undergraduate and graduate students past and present, namely, Aderoke Sakpere, Anesu Marufu, Lwazi Maziya, Nikolai Podlesny, Ammar Abuelgasim, Gregory Harding, Mark Burke, CT Vester, Molulaqhooa Maoyi, Rotondwa Ratshidaho, Sanele Macanda, Dorothy Mhlanga, Lebogang Mametja, Thabo Ndlovu, Nina Otsweleng, Tami Maiwashe, Farzad Motlagh, Md. Hasan Shahriar, Md. Rakibul Islam, Tim Depping, Johannes Klemm, Tomer Yavor, Abrar H. Kamal. You truly have been and remain a source of inspiration, motivation, and continued learning for me.

As a special mention, I would also like to gratefully acknowledge CT Vester, for contributions to Chap. 3; N.J. Podlesny for contributions to Chap. 4; A.B. Sakpere for contributions towards Chap. 5; A.M. Abuelgasim for contributions towards Chap. 6; and D. Reich, R. Islam, and H. Shahriar for contributions towards Chap. 7.

I also gratefully acknowledge funding provided by: (1) National Research Foundation (NRF) South Africa's support for the work on which Chaps. 3, 5, and 6 are based, (2) Hasso-Plattner-Institute for Digital Engineering's support for the work on which Chaps. 1, 2, 4, 7, and 8 are based, and (3) Cimpress GmbH's support for the work on which Chap. 7 is based.

To the members of the Internet Technologies and Systems group, at the Hasso-Plattner-Institute (Potsdam, Germany), working alongside you over the last couple of years has been a pleasure. Thank you for being so supportive and considerate.

Acknowledgements I would like to thank my mentor and adviser, Prof. Dr. Christoph Meinel, for his support in conducting the research on which this monograph is based. His invaluable advice was and continues to be a strong driving force for me. Thank you for pushing me on ever so gently but very firmly.

To my family for their support and patience with me during this time—thank you. Thank you most especially for being patient with me when I was grumpy and for the "alarm clock" throughout the writing journey, and for encouraging me on the bleak and gloomy days.

Over the years I have benefited from the insights and advice of numerous colleagues and mentors. An incomplete list in no particular order includes: Maurice Tchuente, Steve Marsh, Stephen Wolthusen, Karin Frank, Martin Olivier, Robert Hirschfeld, Leonard Barolli, Selim Akl, Patrick Martin, Wendy Powley, Dieter Harms, Berthold Hoffmann, Bernd Kang-Brueckner, Felix Naumann, Andreas Polze, Qiong Feng, Daqing Yang, Vincent Deguin, Andrew Ndumtery, Djamel Benhamou, and Sylvia Osborn.

I would also like to extend a special thank you to my undergraduate and graduate students past and present, namely, Aderonke Sakpere, Anosi Martin, Lwazi Mabizela, Nikolai Podlesny, Ammar Abuelgasim, Gregory Harding, Mark Burke, CJ Vester, Motlatlaphoa Maaye, Rorondwa Katshidaho, Saneie Mlaamba, Dorothy Mohlapa, Lebogang Mametja, Thabo Ndlovu, Nina Otswelelg, Tami Malwiske, Farzad Motlagh, Md. Hasan Shahriar, Md. Rakibul Islam, Tim Depping, Johannes Klemm, Tomer Yavor, Abrar H. Kamal. You truly have been and remain a source of inspiration, motivation, and continued learning for me.

As a special mention, I would also like to gratefully acknowledge C.J. Vester, for contributions in Chap. 3; N.J. Podlesny for contributions in Chap. 4; A.B. Sakpere for contributions towards Chap. 5; A.M. Abuelgasim for contributions towards Chap. 6; and D. Kelch, R. Islam, and H. Shahriar for contributions towards Chap. 7.

I also gratefully acknowledge funding provided by: (1) National Research Foundation (NRF) South Africa's support for the work on which Chaps. 3, 5, and 6 are based, (2) Hasso-Plattner-Institute for Digital Engineering's support for the work on which Chaps. 1, 2, 4, 7, and 8 are based, and (3) Campus Gubh's support for the work on which Chap. 7 is based.

To the members of the Internet Technologies and Systems group at the Hasso-Plattner Institute (Potsdam, Germany), working alongside you over the last couple of years has been a pleasure. Thank you for being so supportive and considerate.

Contents

Part I
Overview

Chapter 1
Introduction

1.1 Overview

In this chapter we provide an introduction outlining and contextualising digital privacy as an allied but independent field of cyber-security. We seek here-in by way of examples, to explain digital privacy solutions cannot be implemented effectively by re-factoring security solutions to emphasise the notion of confidentiality. Primarily, we point out that digital privacy in contrast to security, requires that we reason about the implications of both collective and individual privacy decisions. This as such requires that we consider not only how to keep information secret, but also what the implications of publicly available information are on the ability of systems to protect against disclosure.

1.2 The Paradox of Digital Privacy

For centuries, privacy has been viewed as an individual and subjective concept. This is the sense that what one person might want to keep private, might be something some other person has no problem at all sharing. For instance, it might not matter to Alice that a photograph of her vacationing in the Bahamas was on a newspaper's front page but to Bob this might be a catastrophe. In the physical world therefore the measures one takes to guard ones' privacy are usually, for the most part, geared by individual preferences and sometimes cultural influences.

The digital sphere, and the Internet in particular, requires that we think about privacy as a collective rather than an individual concept. One of the key reasons for this is that in contrast to what happens in the physical world, individual privacy choices on Internet applications have stronger implications for others than they do in the physical world. So while individual privacy decisions and choices can be subjective, in the digital sphere, it is important to consider the impact that individual

A. V. D. M. Kayem, *Enabling Privacy Preserving Data Analytics*, Advances in Information Security 92, https://doi.org/10.1007/978-3-031-93906-8_1

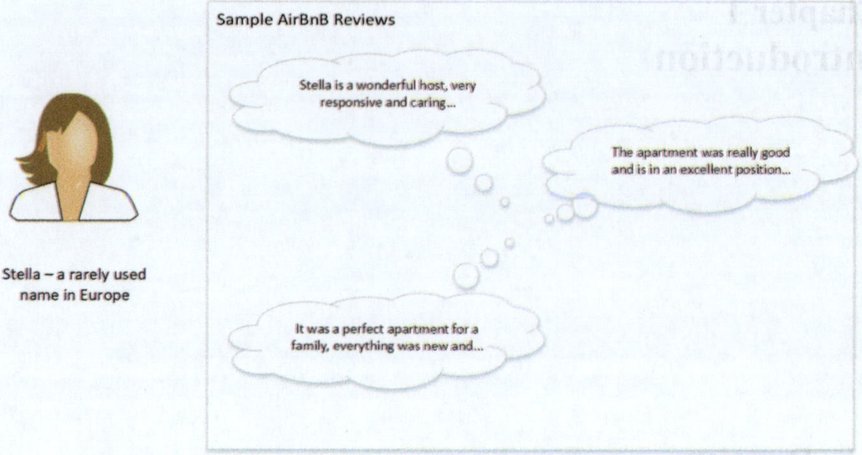

Fig. 1.1 AirBnB data with excerpt of reviews concerning a hypothetical user named "Stella"

Fig. 1.2 Correlating AirBnB data to search engine results

privacy choices have on the collective privacy of the datasets that are generally developed on the basis of user choices.

To put this in perspective, consider the excerpt of data drawn from a publicly available AirBnB dataset given in Fig. 1.1. A closer look at this data indicates that a lady named "Stella" offers AirBnB apartments in the Berlin, Görlitzer park area. As a further step, when this data is combined with search engine data extracted by running a query such as *"Stella AirBnB Görlitzer Park"*, it turns out that "Stella" is a super host who offers 19 apartments in the Berlin area.

So we can infer that Stella is relatively rich. Furthermore, as Fig. 1.2 shows, we also now have her full name "Stella Kaminski". So in this case, the decision to post a review performed by several users had the impact of revealing detailed information about "Stella".

This might not really matter to "Stella" because it serves as unpaid advertising for her. However, if we were to take a look again at Fig. 1.1 and consider that at

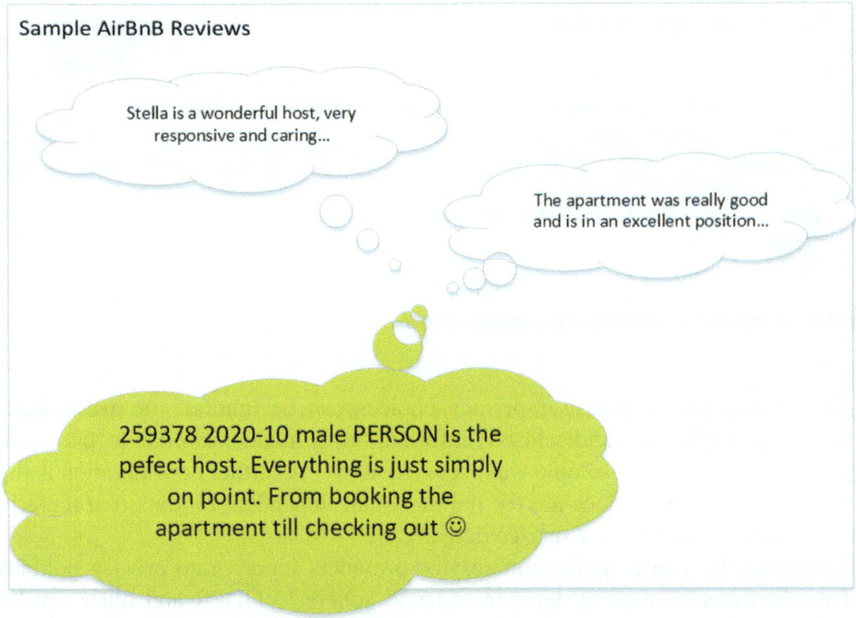

Fig. 1.3 An example of an anonymised record

some stage some form of anonymisation or pseudonymisation were applied to the data, we could for instance obtain a record such as the one depicted in Fig. 1.3.

Again by reverse engineering, we can match terms in the anonymised data version to link back to the reviewer "Calvin" who lives in Amsterdam (see Fig. 1.4). In this case, we see again that "Stella's" privacy choices affect "Calvin's" and vice-versa.

For reasons such as the one we have discussed above, and also because the digital world is increasingly inter-twinned with the physical world, governments have been taking a more proactive approach to developing legislation to regulate usage behaviours in the digital sphere. For instance, the European General Protection Regulation (GDPR) emphasises the importance of data privacy not just as an individual concept but also as a collective concept.

However, implementing mechanisms to guarantee digital privacy is a challenging problem. Particularly when one considers the ubiquity of the Internet and the fact that the growth in processing power is an enabler for the collection, processing, and even generation of very large volumes of complex data. Studies have shown that large volumes of complex personal and sensitive data can be used to predict human behaviours in ways that were hitherto not possible [135, 147]. Technologies such as DeepFake [147] and content sharing on social media platforms[135], highlight ways in which content can be extrapolated and altered to create new content, and also how visual inferences can enable privacy violations.

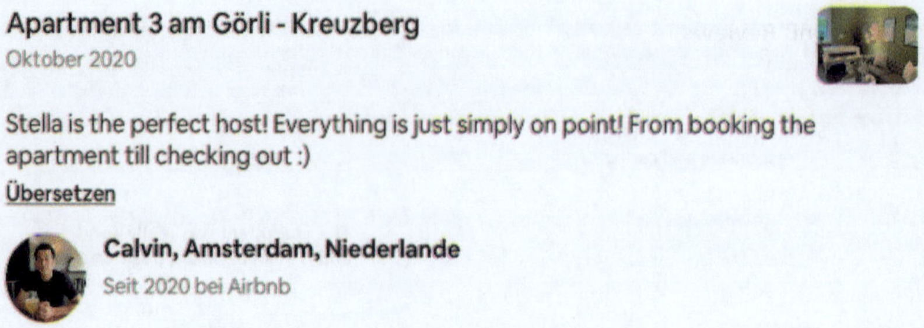

Fig. 1.4 Name of reviewer linked to property owner

The implication is that user privacy choices can be ignored (or overlooked) by inferring hidden (or undisclosed) information, a manner similar to but more sophisticatedly than the example we discussed above. A further consideration is the fact that application services are for the most part dependent on data (particularly personal data) as a driver for service delivery.

Subsequently, while application service providers incorporate privacy policies and mechanisms on their systems in order to adhere to GDPR and other similar privacy regulations worldwide, alternative mechanisms for collecting user data implicitly rather than explicitly, are on the rise [59].

1.3 Digital Privacy as a Collective Concept

In the digital sphere, having mechanisms to ensure data privacy is important both in terms of supporting users in protecting their personal data from exposure and securing the data collected from users. For instance, most security verification and/or authentication mechanisms require some form of personal information to verify that the requesting parties have the authorisations required to access the data, platform, or even application. Yet, at the same time, content sharing platforms such as social media and data intensive applications like the fitness and other health applications residing on IoT[1] devices such as smart phones and sensors, explicitly and implicitly encourage users to share a growing amount private, sensitive, and personal information. These growing volumes of data present a rich source of information to drive data analytics/machine learning services. Therefore, the importance of privacy in the digital sphere underscores the need for supporting such machine learning operations with mechanisms that foster efficient, "correct" and privacy compliant operations. One approach to doing so, and this is the premise of this monograph, is to pre-process the data to ensure that it is privacy compliant before it is submitted to the machine learning model.

[1] IoT: Internet-of-Things.

1.4 Data Pre-Processing: Considerations

Pre-processing data to remove the possibility of linking (or inferences) of sensitive personal information can offer a means of addressing the service providers' need for valid trustworthy data by minimising information loss, while at the same time offering users (data owners) strong guarantees of privacy. While several existing works [150, 151] have shown that anonymised data is prone to privacy disclosures, it is important to underline the fact that context, in terms of the structure of the data and the perceived risks of disclosure, plays a key role in the privacy assertions made about an anonymised dataset.

Therefore, enabling privacy preserving data analytics requires that in addition to transforming data efficiently to minimise the risk of de-anonymisations, that we also take into account the following considerations:

1. The implications data representations have on the choice of transformation mechanisms employed in altering data to achieve privacy. Data representations or the type of data are essential components for anonymisation decisions. Therefore, the type of data in a database or dataset will determine the decision on what the most appropriate anonymisation mechanisms for transforming data should be to achieve privacy. For instance, anonymisation schemes that work well on categorical or numerical data, might not be easily translatable to textual data.
2. The potential use-case scenarios for the data as well as the implications that constraints such as processing power and dataset sizes will have on performance and scalability of the privacy preserving mechanism.
3. The implications of minimising information loss to maintain the veracity of the transformed data while protecting privacy, and what impact this can have on the potential for re-identification. This allows service providers to benefit from the data collected, and to adhere to the privacy requirements of the users (data owners) involved.
4. What measures to adopt in detecting and mitigating of re-identification risks in single and merged datasets in distributed scenarios? This is because the datasets can be collected and/or composed from multiple distributed sources, mechanisms are needed to detect and mitigate risks of de-anonymisation in such datasets while ensuring that the composed datasets adhere to the privacy requirements of the data owners. This applies to re-identification risks in both single and merged datasets in distributed scenarios
5. What approaches to employ in order to deal with bias due to skewed data value representations in generated privacy preserving datasets? Bias due to skewed data value representations can result both in false interpretations of the data but also, as previous work has shown, in de-anonymisations [109, 123, 124, 139]. Therefore, a mechanism to detect and resolve bias risk in anonymised datasets is necessary in ensuring that pre-processed datasets are privacy preserving.
6. The implications for privacy in datasets composed from data belonging to several independent data owners, who might also potentially have subjective opinions about privacy. Accounting for user privacy preferences can serve as a heuristic

for minimising information loss. This however, requires that the implications of the inter-dependencies between individual user privacy preferences are evaluated in terms of inference risks and even potential disclosures of sensitive information.

7. What mechanisms to adopt to nudge data owners or users towards making privacy-aware decisions in sharing personal data or composing privacy preserving datasets. Since the formation of datasets containing personal data is dependent on user data, we must also consider what mechanisms to adopt to nudge data owners or users towards making privacy-aware decision in sharing personal data or composing privacy preserving datasets.

1.5 Problem Statement

Almost every electronic device generates and/or processes data in some form. In fact, statistical records indicate that on average 1.7MB of new data are created per second and that the breach level index reports that about 9,198,580,293 records have been lost or stolen since 2013 [51, 142, 225]. Consequently, analysing these large volumes of data efficiently can make an important contribution to applications driven by data analytics and service delivery. For instance, on online learning platforms being able to analyse student performance data efficiently can yield useful insights on how to improve teaching content, learning materials, and even design software tools to improve learners' experiences. However, analysing the data in a privacy preserving way is an important consideration with respect to existing digital privacy legislation such as GDPR.

Therefore, in this monograph we investigate methods of generating low information loss datasets that are not use-case specific and that account for context-sensitivity without sacrificing privacy. Low information loss ensures that the data retains all or most of the information in the original dataset which is important in supporting query response veracity and as a further step dependable data analytics conclusions. High privacy guarantees that privacy violations are subverted thereby ensuring that the applications using the data operate in a privacy-compliant fashion. Generating low information loss privacy preserving datasets is a performance intensive process, and is known to be NP-Hard [28, 30, 36, 144]. Consequently, having generic privacy preserving data is useful in scenarios where multiple parties need to access the same or similar versions of the data.

In this light and also to address the considerations we mentioned above, our problem statement can be formulated as follows: (1) *"What considerations must be accounted for to generate generic multi-purpose privacy preserving datasets?"* and (2) *"What contextual aspects play an important role in the privacy assertions that are made about the data?"*

By contextual aspects we imply factors that could potentially impede the performance of privacy preserving solutions. For instance, how low processing power impacts on the choice of an anonymisation mechanism, or how data representations impact on the success or failure of an anonymisation mechanism.

1.6 Outline

In this monograph, we present and discuss some approaches to enable privacy preserving data analytics based on the considerations raised in Sect. 1.3 and the problem statement given in Sect. 1.5. We focus on data that appears in the context of many applications such as sensor networks, healthcare applications, security analytics, and social media platforms. In these settings the assumption of temporal continuity and variability is an important consideration in designing privacy preserving data mining (analytics) algorithms. We study the issue of how to transform such datasets in order to retain high utility and high privacy. As mentioned before, high utility is important in supporting the veracity of the results obtained from the data mining/machine learning procedures; while high privacy is a necessary condition in providing strong guarantees of anonymity. Examples of the types of data we discuss in the course of this monograph include mobile data, high-dimensional data, image data, and textual data.

The monograph is sub-divided into three broad topics namely: (1) Causes of de-anonymisation, (2) Methods of Generating Privacy Preserving Datasets and (3) Supporting Safe Privacy Choices.

1.6.1 De-Anonymisation

We review the literature on data anonymisation mechanisms focusing particularly on the weaknesses in these schemes that enable de-anonymisations (Chap. 2). This is aimed at highlighting the challenges that arise in providing strong guarantees of privacy, and to explain why sub-optimal privacy is promoted [163]. We adopt an adversarial approach in that we critique each of the proposed approaches in terms on their vulnerability to disclosures that are exploitable for re-identifications of individuals within a given anonymised dataset. Furthermore, we discuss the role data representations play in enabling de-anonymisation and the challenges that arise in transforming datasets for privacy. Finally, we consider the implications of user data collected from Internet applications such as social media platforms where rich sources of personal information often appear and are very often not well protected. We consider the issues that arise from sharing data on social media platforms and the implications for the designs of security schemes centered around personal identities.

1.6.2 Anonymisation

Under the banner of anonymisations we consider a scenario in which data owners pre-anonymise and outsource their data to a third-party data analytics service provider. Example application scenarios include mobile internet contexts where individual data owners might agree to share data with a data analytics service provider, as well as large data collectors with an obligation to sanitise data to comply with data privacy legislation.

In Chap. 3, we consider a mobile internet context in which individual data owners require a low powered and low processing anonymisation solution in order to generate privacy preserving data that is shared with a third-party service provider such as a data analytics service provider. The datasets generated in this scenario are usually smaller than similar datasets composed on the Internet, and contain fine-grained personal information thereby increasing the risk of re-identification. We present a multi-objective anonymisation solution as a feasible approach to anonymising such datasets effectively and efficiently, by adjusting the trade-off between information loss and privacy to match the use-case scenario.

Composing data from multiple sources generally results in very large high-dimensional data. However, schemes such as multi-objective anonymisation are not well suited to anonymising such high dimensional datasets effectively, and in fact results in high levels of information loss. In Chap. 4, we discuss how this issue can be addressed with a quasi-identifier discovery algorithm. The quasi-identifier discovery algorithm generates feature sets by breaking the functional dependencies that support quasi-identifiers. As such, the feature sets can be published separately without raising a privacy disclosure risk. This presents two benefits, the first is that by breaking the functional dependencies, we remove composite keys from the data and so remove the risk of one-to-one mappings that enable privacy disclosures. Second, generating feature sets implies that the degree of transformations needed to obtain privacy preserving datasets is minimised thereby minimising the level of information loss incurred by the data, while at the same time preserving its usability for data analytics operations.

Data owners typically anonymise their data based on some pre-defined set of privacy preferences or inline with an in-house privacy policy. Generating privacy preserving datasets based on compositions of these individual datasets requires that we give consideration to privacy as both a subjective and collective concept. In Chap. 5, we discuss the implications of taking into consideration user privacy preferences in generating privacy preserving data and discuss some approaches to integrating individual data owner privacy choices/preferences in generating privacy preserving datasets.

1.6.3 Safe Privacy Choices

In Chap. 6, in order to account for the fact that most of the data collected on digital platforms is contributed either directly or indirectly by users, we present one mechanism for supporting safe privacy choices and data sharing on social media platforms. To this end, we present a mechanism for accounting for user privacy preferences in generating anonymised data. This is in a bid to consider instances in which sub-optimal privacy might be an acceptable compromise in order to minimise information loss. The proposed mechanism uses nudges or recommendations to enable safer user-conscious privacy decisions. Mechanisms such as these can serve as a step for supporting data analytics mechanisms that operate on shared or co-owned content. In this case, knowledge about a user's privacy preferences or choices is helpful in transforming content to protect individual privacy choices while at the same time supporting privacy preserving data analytics.

Nudging users to make safer privacy decisions naturally raises the question of how to deal with individual privacy decisions, the subjectivity of those decisions, and their impact on collective privacy assertions. As such, in Chap. 7, we discuss the considerations that need to accounted for in discovering personal identifying information in data, and the privacy implications of transforming the data. In particular, we discuss how techniques such as outlier detection, bayesian inference, and deep learning could potentially offer a means of analysing datasets on-the-fly to check for privacy disclosure risks/vulnerabilities.

We conclude in Chap. 8, with a summary of the contributions of this monograph, and a discussion of potential avenues for future work.

Part II
Data De-Anonymisation

Chapter 2
De-Anonymisation Mechanisms: An Overview

2.1 Overview

Preserving the privacy of users on web environments is a key requirement for analytics applications such as recommender systems, crowdsourced platforms, and web-scale data mining or machine learning applications. Yet, in spite of the existence of best effort anonymisation algorithms, incidents of privacy breaches remain a concern. In this chapter, we present existing mechanisms for transforming data to obtain usable and privacy preserving datasets both from the syntactic data transformation perspective. We focus on the vulnerabilities of these data transformation mechanisms to privacy breaches, in an effort to explain why and how privacy exploits occur. For simplicity, the examples we use to illustrate privacy vulnerability characterisations are based on a relational database model but can be extended to other data models or representations.

As an added dimension and also for completeness we discuss data collection models and platforms, and the privacy implications of these models on data compositions in distributed environments such as the Internet.

We wrap up the chapter with a discussion on the theoretical performance implications for data transformation algorithms, specifically in terms of generating anonymised data, in a bid to underline why most existing anonymisation algorithms can only guarantee sub-optimal privacy.

2.2 Achieving Privacy

Achieving data privacy is conditional on some form of distortion of the original data. In distorting the data however, we must aim to preserve its utility. Data utility is an important consideration when the transformed data is used for computing operations that require truthful query results without putting privacy at risk. Examples arise

A. V. D. M. Kayem, *Enabling Privacy Preserving Data Analytics*, Advances in Information Security 92, https://doi.org/10.1007/978-3-031-93906-8_2

in privacy preserving machine learning, data analytics, market analysis, and health analytics. In these cases, a high degree of data utility can help ensure veracity of the results. That is the same answer to a query can be provided irrespective of whether or it was run on un-transformed or transformed data.

There are two tenets to addressing the issue of generating privacy preserving data. In the first case, the data is transformed to ensure privacy but minimise information loss. In general, anonymisation mechanisms that rely on data transformation algorithms to minimise information loss are referred to as *syntactic data anonymisation* mechanisms. This is because the anonymisation is achieved by syntactically modifying the data without accounting for the semantics (or the meanings) drawn from the published data. Some examples of *syntactic data anonymisation* mechanisms include k-anonymity, l-diversity, and t-closeness which we discuss later on in this Chapter. The second case advocates probabilistically distorting the data on a per-query basis to provide truthful query responses without putting privacy at risk [67]. Data anonymisation mechanisms supported by this principle are typically referred to as *semantic data anonymisation* mechanisms. A classic example of this is *Differential Privacy* [65, 66].

Providing firm guarantees of data utility however, comes at the price of privacy. That is, when the degree of data utility is high, concealing private and sensitive individual user information is a challenge. A further issue is that most existing algorithms for anonymising data have been shown to be NP-hard [27, 62, 127, 144, 238], therefore these mechanisms can only guarantee sub-optimal privacy [163].

Generating useful privacy preserving data requires therefore that the transformation procedure be designed to ensure a cost-benefit tradeoff between usability and information loss. In the following Section we consider one of the first (and still widely used) approaches to transforming data for privacy.

2.3 Pseudonymisation

Data pseudonymisation was one of the first methods used to used to transform data for privacy. With pseudonymisation, explicit identifying attribute-values are replaced by pseudonyms. The idea is that by replacing the original attribute values with pseudonyms, it should be impossible to guess or infer the exact original value. For instance, in Table 2.1 we have some data on adults and the corresponding diseases that these adults are afflicted with. Table 2.1 contains some **explicit identifiers** which are basically attributes that can be used to directly identify individuals. In this case, two examples of explicit identifiers in Table 2.1 are "ID-Number" and "Name".

The main advantage of pseudonymisation is that if needed, the original values can be obtained by a reverse transformation of the pseudonym. Using pseudonymisation, one example of what we could do to eliminate identifying information is to replace the explicit identifier attribute values with pseudonymous ones. In practice, more complex approaches can be used to obtain pseudonyms such as by means of one-

Table 2.1 Original dataset: Adults and corresponding diseases

Name	ID-Number	Age	Gender	Postcode	Disease
Alice	1827-88	34	Female	34521	COVID-19
Bob	1727-88	32	Male	34530	COVID-19
John	1526-98	33	Male	34540	COVID-19
Tina	5555-88	37	Female	34551	Lung cancer
Jane	6654-78	38	Female	34570	Gallstones
Joyce	2345-66	36	Female	34580	Malaria
John	7896-54	42	Male	34555	Thyphoid
Jamie	6789-23	43	Male	34560	Lung cancer
James	8967-23	44	Male	34525	Stomach cancer

Table 2.2 Original dataset with pseudonymous values in replacement of personal identifier values

Name	ID-Number	Age	Gender	Postcode	Disease
xyz	182-XXX	34	Female	34521	COVID-19
ykt	172-XXX	32	Male	34530	COVID-19
txc	152-XXX	33	Male	34540	COVID-19
wdb	555-XXX	37	Female	34551	Lung cancer
bet	665-XXX	38	Female	34570	Gallstones
xxx	234-XXX	36	Female	34580	Malaria
des	789-XXX	42	Male	34555	Thyphoid
dbs	678-XXX	43	Male	34560	Lung cancer
kjl	896-XXX	44	Male	34525	Stomach cancer

way hash functions [120], and taxonomies [131]. For simplicity however, in the example given in Table 2.2, we assume that pseudonyms are assigned arbitrarily. So for instance, "Alice" (Name) and "1827-88" (ID-Number) are replaced by "xyz" (Name) and "1827-XXX" (ID-Number) respectively.

2.4 Sweeney's De-Anonymisation

Sweeney however, demonstrated that using pseudonyms does not guarantee privacy. In fact, Sweeny demonstrated that by combining pseudonymised US Census data with auxiliary data such as health data and/or voter registration data, it is possible to uniquely identify up to 87% of the US population based on a combination of attributes such as *gender, ZIP Code,* and *full date of birth* [210]. To provide an illustration of how this might occur, we consider Table 2.2, and consider how it might be combined with a hypothetical voter registration dataset such as the one given in Table 2.3.

Table 2.3 Voter registration data

Name	Birth date	Gender	Postcode	Political opinion
Alice	18 Sept. 1987	Female	34521	Democrat
Bob	17 Jun. 1989	Male	34530	Liberal
Jeremy	16 Jul. 1988	Male	34540	Green
John	12. Aug. 1988	Male	34551	Liberal
Jamie	14. June 1982	Male	34570	Green
James	10 May 19	Male	34580	Democrat

A closer look at both Tables 2.2 and 2.3 indicates that it is possible to deduce, for instance, that the first two records in both datasets (Tables 2.2 and 2.3) correspond to Alice and Bob respectively. Additionally, more sensitive information such as the *political opinions* and *disease afflictions* of both individuals can be inferred.

In a subsequent result, Phillipe Golle [93], corroborated Sweeney's result but asserted that disclosing one's *birth date, zipcode,* and *gender* resulted in 63% unique identifications in the US population. Both results however, agree that pseudonymised data is not an adequate approach to data privacy.

As a side note, other well cited examples of re-identifications of individuals based on poorly pseudonymised data include: (1) The AOL dataset and the re-identification of Thelma Arnold [24]; and (2) The re-identification of Governor Weld [211].

We now discuss Sweeney's solution to the issue of alleviating disclosure risks due to pseudonymised data.

2.5 *k*-Anonymity

Sweeney [210] and Golle [93] pointed out that, in addition to explicit identifiers, combinations of attributes can be used to re-identify an individual. Such attribute combinations are referred to as quasi-identifiers [52]. Quasi-identifiers can be defined more formally as follows.

Definition 2.1 (Quasi-Identifiers) Given a table T that is defined by a set of attributes $A = A_1, \ldots, A_M$, a quasi-identifier Q is a set of attributes $Q = A_i, \ldots A_w$ such that $1 \leq i, w \leq M$ and $Q \subseteq A$, that uniquely re-identifies (or provides a one-to-one mapping) at least one tuple $t_i \in T$.

Example 2.1 Table 2.2 provides several examples of quasi-identifier attributes such as {Age, gender, Postcode}.

In order to guard against re-identifications on pseudonymous data, due to inferences drawn from quasi-identifiers, Sweeney proposed k-anonymity as a model for protecting privacy [213, 214]. k-anonymity, classifies each tuple (t_i) with at least $k-1$ similar data tuples, to ensure indistinguishability. These cluster classifications are typically termed **Equivalence Classes**. An equivalence class can be defined informally as follows:

Definition 2.2 (Equivalence Class) An equivalence class is a set or group of records that are similar in terms of a shared or common quasi-identifier value(s).

Example 2.2 Table 2.5 provides examples of three (3) equivalence classes built around the quasi-identifier *(Age, Gender, & Postcode)*. In this case, R1, R2, and R3 form an equivalence class based on similarity of the quasi-identifier (Age = [31–35], Gender = Adult, and Postcode = 345**).

For example in Table 2.5, we have 3 equivalence classes of size 3 (i.e. $k = 3$), generalisation was applied to the Age and Gender attributes, and suppression to the last two (2) digits of the postcode attribute. A further step would be to completely suppress the 'Name' attribute values since these explicitly identify individuals in the dataset.

The $k - 1$ nearest neighbours are selected based on similar quasi-identifier attribute-values and is enforced by generalisation and suppression [213].

2.5.1 Generalisation, Suppression, and Perturbation

As explained before, de-identifying data by removing all explicit identifiers as is the case in pseudonymisation, is not enough to guarantee privacy. To generate a privacy preserving dataset therefore, a first step is to de-identify the data. We define de-identified data as follows:

Definition 2.3 (De-Identified Data) This is data generated as a result of a series of transformation procedures (e.g. Generalisation and Suppression) that identify and replace or remove explicit identifiers from a dataset. Replacement procedures can involve changing the original value to some alternative that is created based on a well-defined taxonomy or ontology, or a pseudonymous value.

Example 2.3 Table 2.2 provides two examples of how explicit attribute values are replaced or modified for pseudonymous values.

De-identifying data requires that, in addition to creating pseudonyms for explicit attribute-values, transformation algorithms such as *generalisation* [193, 209, 213], *suppression* [192], and *perturbation*[184] are applied to the data to minimise disclosure risks. Explicit identifiers can be defined as follows:

Definition 2.4 (Explicit Identifiers) Let $E = \{E_1, \ldots, E_x\}$ where $E \subset A \in T$ be a set of explicit identifiers and let $|U|$ be the cardinality of unique attribute values in a given explicit identifier, E_i. An attribute E_i is defined as one in which $|U| \approx N$. That is, we can formulate an injective function $f(E_I)$ such that the values of E_i result in one-to-one mappings on all or most of the tuples in T.

Table 2.4 User data

ID	Phone	Name	Birth date	Gender	Postcode
812	0123456	Alice	1987	Female	3452
978	0995676	Bob	1989	Male	3453
3456	0167832	Jeremy	1988	Male	3454
234	0978654	John	1988	Male	3455
742	0152345	Jamie	1982	Male	3457
654	0134256	James	1985	Male	3458

for tree= font=, line width=1pt, draw, child anchor=north, parent anchor=south, grow=south, align=center, edge path= [line width=1pt, edge] (!u.parent anchor) –- ([yshift=4pt].child anchor) – (.child anchor) edge label; , , s sep+=20pt, [Human, fill=white,double=white, [Adult,double=white [Male] [Female]] [Teenager,double=white [Male] [Female]]]

Fig. 2.1 Example of a generalisation taxonomy

Example 2.4 In Table 2.4 attributes such as *ID-Number, Name, Phone-Number,*... that without any extra information allow for the unique identification of a designated individual can be qualified as *Explicit Identifiers*. Notice in this case, that an attribute that qualifies as a PII is also an explicit identifier, but depending on the degree of the cardinality of unique value occurrences, an explicit identifier does not necessarily qualify as personal identifying information (PII).

Generalisation algorithms operate by applying values drawn from a pre-defined taxonomy as replacements in order to transform the data for privacy [193, 213, 231, 234]. For example, using the taxonomy given in Fig. 2.1, 'Male' and 'Female' might be generalised to 'Adult' for all persons over 18, and for further generalisation to 'Human'. Each change progressively increases the degree of information loss due to the distortions (transformations) while augmenting privacy.

Definition 2.5 (Domain Generalisation) A domain $D^* = \{P_1, P_2, \ldots\}$ is a generalization (partition) of a domain D if $\bigcup P_i = D$ and $P_i \cap P_j = \emptyset$ whenever $i \neq j$. For $x \in D$ we let $\theta_{D^*}(x)$ denote the element $P \in D^*$ that contains x.

We can as such create a partial order $\prec_G$ on domains by requiring that $D \prec_G D^*$ if and only if D^* is a generalisation of D. This is in line with the representation we have in Fig. 2.1. In this case, the generalisation of the gender attributes 'Male' and 'Female', can depending on contextual information such as 'Age' get transformed into 'Adult' or 'Teenager'; or as a further step to 'Human'.

Suppression works by deleting parts of or whole data values to conceal the original value, while perturbation moves (swaps) attribute-values between tuples in the dataset [192]. A primary goal of perturbation is to distort the data to retain its semantic meaning while at the same time preserving privacy [184]. However, the distortions are aimed at reducing the levels of information loss by avoiding changes to the attribute-values. Instead, tuples and attribute-values are moved to enable equivalence class formations. When this is done repeatedly and for large datasets, the swapping operations can become performance intensive. Furthermore, swapping (perturbations) can result in highly inaccurate data which in turn is useless for data analytics operations.

To a certain extent, high information loss can be addressed by perturbation. Perturbation operates by modifying the original un-anonymised attribute-value to the closest similar value. In order to achieve this, an aggregated value or an approximation value is used instead of generalisation or suppression. This is done on a per-case basis and so reduces information loss by instead modifying only one (or in the extreme case only a few values) to enable equivalence class formation. A further step is to move the affected tuple or attribute-values to an equivalence class that hides the value better without requiring a change in its content. However, in large datasets, efficiently replacing and/or moving attribute-values (or tuples) is a time consuming and processing intensive operation [78, 220, 221].

Finally, pertubation replaces original values with synthetic ones, so that statistical conclusions or information drawn from the data is not very different from the original. Examples of approaches that employ perturbation as a primary means of generating privacy preserving datasets include differential privacy where techniques such as additive noise are used to modify data values for concealment [65, 66].

Other techniques includes ones such as data swapping, permutations, and anatomisation which Fung et al. [82] discuss in depth. However, for the purposes of this monograph, and for simplicity we focus primarily on how generalisation and suppression are applied to generate privacy preserving datasets. To this end, in the following Section we discuss how generalisation and suppression are employed to enable the generation of k-anonymous privacy preserving data.

2.5.2 The k-Anonymisation Scheme

As mentioned before, in order to enforce privacy by indistinguishability, k-anonymity classifies each tuple with $k-1$ other tuples based on quasi-identifier value similarity. In contrast to pseudonymisation therefore, k-anonymisation seeks to ensure that privacy violations are not possible even when explicit identifiers are removed.

For simplicity, in the examples that follow, we assume that there is only one sensitive attribute. However, our discussions can easily be extended to scenarios involving multiple sensitive attributes. The attributes that are not classified as sensitive attributes are classified as non-sensitive.

Let $T = \{t_1, t_2, \ldots, t_N\}$ be a table with attributes $A = \{A_1, \ldots, A_M\}$ where T is a subset of some larger dataset D such that each tuple $t_i \in T$ represents a tuple (record) describing an individual within D; and T is represented by an $N \times M$ matrix of attributes and tuples. In T, $t[A_i]$ denotes the attribute-value of A_i for the tuple t, and A_i is described by a set of values $v_1, \ldots, v_N$.

Definition 2.6 (Sensitive Attributes) Let $S = \{S_1, \ldots, S_k\}$ be a set of all sensitive attributes such that $S \subset A$. An attribute A_i is described as a *sensitive attribute*, S_i if and only if S_i is an attribute whose values must be kept secret, from parties that have no direct access to the original data.

Example 2.5 The hospital records on disease given in Table 2.1 and the voters' political opinions in Table 2.3 provide examples of the fact that the associations between a person and their medical condition or political opinion should be kept secret.

Sensitive attributes are often associated with the concept of personal identifying information (PII), in that personal identifying information (PII) is described as data whose granularity in terms of characterising attribute-values, can be used to identify an individual uniquely. For example, in Table 2.1 the names and ID-Numbers serve as examples of personal identifying information (PII). In this case, each attribute-value contains information related to a unique individual.

Non-sensitive attributes include two kinds of attributes namely: *Explicit Identifiers* and *Quasi-Identifiers*. *Explicit Identifiers* share some similarity with personal identifying information (PII) in that they are considered to be attributes containing a high number of unique values and so have an increased potential of being used for one-to-one re-identifications of individuals in a dataset.

More formally, k-anonymity can be defined as follows:

Definition 2.7 (k-anonymity) A table T satisfies k-anonymity if, for every tuple $t_i \in T$ there exist $k-1$ other tuples $t_{i_1}, t_{i_2}, \ldots, t_{i_k-1} \in T$ such that $t[C] = t_{i_1}[C] = t_{i_2}[C] = \ldots = t_{i_k-1}[C]$ for all $C \in QI$ where $C = C_1, C_2, \ldots, C_p \subseteq A$, $t[C]$ denotes the tuple $(t[C_1], \ldots, t[C_p])$ which is a projection of t onto the attributes in C and QI is the set of quasi-identifiers.

Example 2.6 Table 2.5 provides an example of a k-anonymised table based on similar values for the quasi-identifier attributes {Age, Gender, Postcode}. In this case, we have 3 equivalence classes each of size $k = 3$.

Table 2.5 Dataset: 3-equivalence classes, generalisation on *age & gender*, and suppression on *postcode* attribute-values

Record #	Age	Gender	Postcode	Disease
R1	[31–35]	Adult	345**	COVID-19
R2	[31–35]	Adult	345**	COVID-19
R3	[31–35]	Adult	345**	COVID-19
R4	[36–40]	Female	345**	Lung cancer
R5	[36–40]	Female	345**	Gallstones
R6	[36–40]	Female	345**	Malaria
R7	[41–45]	Adult	345**	Thyphoid
R8	[41–45]	Adult	345**	Lung cancer
R9	[41–45]	Adult	345**	Stomach cancer

2.5.3 Discussion

It should be noted however that adversarial background knowledge about the generalisation schema used, knowledge of the existence of some individuals in the dataset, and knowledge of the distributions of sensitive and non-sensitive values such as demographic background knowledge [139, 232, 233, 235], can play an effective contributing role in re-identification attacks on k-anonymised data. We discuss this in more depth in Sect. 2.9, but in the following Section we focus on how k-anonymised data can be de-anonymised.

2.5.4 Attacks to k-Anonymised Data

Due to its conceptual simplicity k-anonymity is popularly used to anonymise data. Examples of anonymisation toolkits that offer algorithms to k-anonymise data include: ARX [179] and eSito[74]. However, as Machanavajjhala et al. [139] showed, k-anonymised datasets are vulnerable to homogeneity and background knowledge attacks. In addition, Sweeney[212] also discussed the vulnerability of k-anonymised data to attacks such as unsorted matching, complementary release, and temporal disclosures. We discuss each in more depth below.

2.5.4.1 Homogeneity Attacks

Homogeneity attacks take advantage of homogeneous distributions of attribute values to infer hidden patterns within the dataset. These hidden patterns can then be built upon to re-identify one or several individuals within the dataset. For instance, in Table 2.6 suppose Alice and Bob are neighbours and that Bob were to fall seriously ill and be taken to hospital by an ambulance. Alice is curious to find out if Bob was taken to hospital because he is infected with the COVID-19 virus. Hypothetically, what she could do is to check, Table 2.6 which she might have come across in searching the Internet for health datasets published by COVID-19 tracing applications, as well as nearby hospitals for current inpatients. Based on her knowledge that Bob's records must be in Table 2.5, Alice starts looking for clues to confirm her hypothesis. First she knows that Bob is male, but discounts the last three (3) records in Table 2.6 because she also knows that Bob is 32 years old. Second, she guesses that records, $R12$, $R23$, $R34$, $R78$, $R89$, $R99$ might have been generalised on the "Gender" attribute, to conceal and uneven distribution of values because all the affected individuals live in a Postcode radius of "345**". So, she comes up with the following conclusion.

- Bob's record is one of records 1–3, because Bob is in his early 30s. Furthermore, since every one of records 1–3 is associated to COVID-19, Bob must have been taken to hospital because of this.

Table 2.6 Dataset: 3-equivalence classes, generalisation on *age & gender*, and suppression on *postcode* attribute-values

Trace ID #	Age	Gender	Postcode	Disease
R12	[31–35]	Adult	345**	COVID-19
R23	[31–35]	Adult	345**	COVID-19
R34	[31–35]	Adult	345**	COVID-19
R45	[36–40]	Female	345**	Lung cancer
R56	[36–40]	Female	345**	Gallstones
R67	[36–40]	Female	345**	Malaria
R78	[41–45]	Adult	345**	Thyphoid
R89	[41–45]	Adult	345**	Lung cancer
R99	[41–45]	Adult	345**	Stomach cancer

Table 2.7 Background information on John

Name	Age	Gender	Postcode
John	42	Male	34555

Alice's de-anonymisation of Bob's data is due in part to the fact that the disease attribute values are all the same. That is, homogeneity attacks are enabled by the homogeneity of attribute-values.

The possibility of homogeneity attacks was discussed in depth by Ohrn and Ohrn-Machado [164] and Machanavajjhala et al. [139]. Ohrn and Ohrn-Machado proposed a solution to address homogeneity attacks, which is similar to *entropy l-diversity* that Machanavajjhala et al. [139] proposed.

2.5.4.2 Background Knowledge

The background knowledge attack extends the homogeneity attack by employing information that the adversary knows about the victim and some auxiliary data. In the case of background attacks to *k*-anonymous data, consider again Table 2.6 and specifically the last three records—R78, R89, and R99.

To conduct the attack, Alice (the adversary) uses verified background knowledge that she has on John in terms of background information (see Table 2.7) to determine if John's data is within the anonymised dataset, and second, what type of illness John might be suffering from.

Looking R45, R56, and R67, Alice can eliminate these records as a possibility on the basis that all the candidates are "Female", which leaves R12, R23, R34, R78, R89, and R99 for consideration. Again, due to the information Alice has from Table 2.7, she can also eliminate R12, R23, and R34 from consideration on the basis of "Age", which leaves R78, R89, and R99. In the equivalence class containing R78, R89, and R99, Alice figures out that the term "Adult" was used to generalise the gender attribute-values to conceal the fact that one of the values is either "Female" or "Male", and that the other two (2) are monotonic, that is both either "Female" or "Male". She then looks at the other attributes—Postcode is similar across the equivalence class, i.e. with the prefix "345–", and the same with the disease attribute. Therefore, she concludes that John has a probability of $\frac{1}{2}$ of having either "Cancer" or "Thyphoid". Furthermore, if from auxiliary information Alice knows that John is not suffering from "Cancer", then she conclude with a probability of 1 that John has Thyphoid.

2.5.4.3 Unsorted Matching Attack

The unsorted matching attack takes advantage of the order in which tuples appear in the released table. We recall that homogeneity and background attacks also leverage this same characteristic in published anonymised datasets. For example, in the case of the background attack described in the previous section, if the positions of the tuples within the anonymised table were to have been randomly determined (e.g. via perturbation), the job of the adversary (Alice) would have been made harder. However, as we have mentioned before, re-ordering the tuples in the dataset can in certain cases negatively affect the accuracy of results of data analytics operations on the data [45, 129].

2.5.4.4 Complementary Release Attack

Complementary release attacks take advantage of historical releases of data. In this case, the attributes that compose the data in the released datasets could be viewed as quasi-identifiers within newer versions of anonymised datasets that are composed from the original dataset. An example of a complementary release attack could occur If two tables say Tables 2.8 and 2.9, are published one after the other.

In both tables—Tables 2.8 and 2.9—on studying records R1–R3 (highlighted in both tables), it can be observed that the data released in Table 2.9 provides more detail on the exact "Age" and more detail on the "Postcode". So by releasing Table 2.9, the privacy provided by Table 2.8 is put at risk.

Table 2.8 Dataset: 2-equivalence classes, quasi-identifier = *age, gender, postcode*

Record #	Age	Gender	Postcode	Disease
R1	[31–35]	Adult	345**	COVID-19
R2	[31–35]	Adult	345**	COVID-19
R3	[31–35]	Adult	345**	COVID-19
R4	[36–40]	Female	345**	Lung cancer
R5	[36–40]	Female	345**	Gallstones
R6	[36–40]	Female	345**	Malaria

Table 2.9 Dataset: 2-equivalence classes, quasi-identifier = *birthdate, gender, postcode*

Record #	Age	Gender	Postcode	Disease
R1	1987	Adult	34520	COVID-19
R2	1987	Adult	34520	COVID-19
R3	1987	Adult	34520	COVID-19
R4	1982	Female	34550	Lung cancer
R5	1982	Female	34550	Gallstones
R6	1982	Female	34550	Malaria

Table 2.10 Dataset: 2-equivalence classes, quasi-identifier = *birthdate, gender, postcode*

Record #	Birthdate	Gender	Postcode	Disease
R1	1987	Adult	345**	COVID-19
R2	1987	Adult	345**	COVID-19
R3	1987	Adult	345**	COVID-19
R4	1982	Female	345**	Lung cancer
R5	1982	Female	345**	Gallstones
R6	1982	Female	345**	Malaria

Table 2.11 Dataset: 2-equivalence classes, quasi-identifier = *birthdate, gender, postcode*

Record #	Birthdate	Gender	Postcode	Disease
R1	1987	Male	34520	COVID-19
R2	1987	Female	34520	COVID-19
R3	1987	Female	34520	COVID-19
R4	1982–85	Female	34550	Lung cancer
R5	1982–85	Female	34550	Gallstones
R6	1982–85	Female	34550	Malaria

2.5.4.5 Temporal Attack

Temporal attacks build on the notions that enable both unsorted match and complementary release attacks, in the sense that data releases are dynamic and also to a large degree conditional on the use case. Typically data can change, either by the addition or insertion of records, or even changes in quasi-identifier values. For instance, if we consider Tables 2.10 and 2.11, and assume that Table 2.10 is released at time t_1 and Table 2.11 is released at a later time, say t_2. In Table 2.10 we have a quasi identifier of *Birthdate, Gender, and Postcode* and equivalence classes of size 3 (i.e. $k = 3$, by k-anonymity).

While Table 2.11 contains the same quasi-identifier of *Birthdate, Gender, and Postcode*. We notice as indicated by the highlighted attribute-values that combining the two tables (i.e. Tables 2.10 and 2.11) and linking based on the disease attribute, an adversarial party can obtain the following extra pieces of information: the real postcodes of R4–R6; the actual gender for R1–R3; and the actual year of birth for R4–R6. In addition, it becomes evident that the postcode area 34520 is heavily affected by COVID-19.

We now look at l-diversity as a method, that has been proposed for dealing with attacks to k-anonymous data.

2.6 l-Diversity

Machanavajjhala et al. [139] proposed addressing the suite of attacks against k-anonymised datasets, with the notion of l-diversity. Basically, l-diversity works by considering the diversity of distributions of sensitive attribute-values within an equivalence class with respect to the associated quasi-identifiers.

The l-diversity approach proposes the notion of *bayes-optimal privacy* as means of modelling background knowledge. Bayes-optimal privacy relies on Bayesian inference to determine whether or not a given distribution of sensitive attribute-values poses a privacy violation risk. Both *bayes-optimal privacy* and Bayesian inference techniques are used as a the basis for calculating the probability of *positive disclosure* or *negative disclosure* based on an adversary's background knowledge with respect to a published anonymised table T^*.

2.6.1 *Positive Disclosure*

Positive disclosure, implies that the distribution of sensitive attribute-values in T^* is such that there is a high privacy violation risk. To exploit the violation risk the adversary must have access to background knowledge B_T that allows him/her to compute with a high probability, the existence of a given tuple t_i in T^*. In addition, the adversary must be able to confirm with a high degree of certainty that there is an exact match between t_i and $t_i' \in B_T$, and that this match enables the deduction of private information regarding t_i. Homogeneity attacks (see Sect. 2.5.4.1) are an example of positive disclosures.

2.6.2 *Negative Disclosure*

By contrast, *negative disclosure* occurs if the adversary can correctly exclude certain values and or tuples with a high probability and in this way deduce the existence or non-existence of a given tuple in an anonymised table T^*. For example in Table 2.11, a negative disclosure occurs if we can correctly deduce that a given individual associated to t_i, say Alex, is not in the dataset and does not have any of the diseases listed.

2.6.3 *Uninformative Principle*

As a first step, $l-$diversity employs the *Uninformative Principle* where by the published table T^* is structured such that it limits the amount of extra information that the adversary can get access to. That is there should be no difference between prior and posterior knowledge of the adversary. Evfimievski et al. [76] provide a definition of privacy breaches in line with the uninformative principle. The idea is that given an anonymised table T^*, bayes optimal privacy is met if and only if the transformations applied to T to obtain T^* are such that observations of T^* do not provide any information about T. In fact T and T^* should appear to be similar when observed or studied. In addition to background knowledge, bayes optimal privacy also models diversity. Diversity seeks to reduce the uniformity of sensitive attribute-values to reduce the disclosure risk.

However, bayes-optimal privacy poses a number of limitations in terms of its practical application to real-world datasets. The first is that the data publisher generally has little or no knowledge of the sensitive attribute-value distributions for the general population. So, making privacy assertions about the distributions of values in the sample represented in the published anonymised datasets is a challenging problem. Furthermore, the data publisher cannot determine, beforehand how much and what background information an adversary might have.

For instance, in Table 2.11, the data publisher does not know how much background knowledge Alice has on John. Thirdly, the bayes optimal privacy model assumes that knowledge can always be modelled probabilistically whereas in practice this is not the case. For example, John's daughter might tell Alice innocently that he tested negative for COVID-19. Fourthly, the assumption is that the data set only has to deal with one adversary at a time, whereas in practice adversaries might coordinate their actions to improve their success rates and also to avoid conducting several false positive searches.

2.6.4 Alleviating the Drawbacks of Bayes Optimal Privacy

The l-diversity principle addresses the weaknesses of bayes-optimal privacy by proposing that each anonymised dataset is structured such that there are at least l distinct sensitive attribute-values in each equivalence class. For example, if we consider again Table 2.11, the equivalence class containing R1–R3 is not l-diverse, where as R4–R6 is l-diverse. If we were to modify Table 2.11 to obtain Table 2.12, the attacks against the 3-anonymous table (see Table 2.11) can be prevented by modifying the distribution of sensitive attribute-values to obtain the 2-diverse table given in Table 2.12. Alice can no longer infer, with a probability of 1, that Bob (a 31 year old, male from postcode 34520) has COVID-19 (see Table 2.12).

To deal with issues centered around information gain and the potential likelihood of disclosure, Machanavajjhala et al. [139] also offer two instantiations of the l-diversity principle namely entropy l-diversity based on the information theoretic notion of entropy, and recursive (c, l)- diversity which considers the number of occurrences of the ith most frequent sensitive attribute value with in an equivalence class. Other l-diverse measures include, distinct and probabilistic l-diversity which both consider distributions of values in terms of frequency of occurrence or

Table 2.12 2-diverse dataset based on the sensitive attribute "disease"

Record #	Birthdate	Gender	Postcode	Disease
R1	1982–1990	Adult	345**	Gallstones
R2	1982–1990	Adult	345**	COVID-19
R3	1982–1990	Adult	345**	COVID-19
R4	1982–1990	Adult	345**	Lung cancer
R5	1982–1990	Adult	345**	COVID-19
R6	1982–1990	Adult	345**	Malaria

Table 2.13 2-diverse dataset based on the sensitive attribute "disease" with a 4-anonymous dataset

Record #	Birthdate	Gender	Postcode	Disease
R1	1982–1990	Adult	345**	Gallstones
R2	1982–1990	Adult	345**	COVID-19
R3	1982–1990	Adult	345**	COVID-19
R4	1982–1990	Adult	345**	Stomach cancer
R5	1982–1990	Adult	345**	Lung cancer
R6	1982–1990	Adult	345**	COVID-19
R7	1982–1990	Adult	345**	Malaria
R8	1982–1990	Adult	345**	Malaria

probability of inference based on observations of the distributions of the values. In practice, both versions can be restrictive performance wise and so a cost-benefit tradeoff is needed to find the optimal level of privacy by l-diversity [124].

In the next Section we discuss attacks to which l-diverse privacy preserving datasets are potentially vulnerable.

2.7 Attacks to l-Diversity

Datasets that are anonymised using the principle of l-diversity, can in certain cases be vulnerable to skewness and similarity attacks [123, 124]. As an example of how skewed sensitive attribute-value distributions can be exploited to violate the privacy guarantees of l-diverse anonymised datasets, consider again the 2-diverse table given in Table 2.12) and a modified version given in Table 2.13.

Table 2.13 is distinct 2-diverse, entropy 2-diverse, as well as recursive $(c, 2)$-diverse. However, a closer look at the distribution of sensitive values indicates that in equivalence class#1, a person has a 50% chance of being COVID-19 positive, and in equivalence class#2, a 50% chance of having malaria. This presents a privacy disclosure risk, because Alice could for instance, based on the post-code guess that the probability of Bob having malaria is much lower than it being COVID-19 and so focus her de-anonymisation efforts on equivalence class#1 instead of both. Furthermore if equivalence class#1 has 3 records with COVID-19 as a sensitive

Table 2.14 2-diverse dataset based on the sensitive attribute "disease" with a 4-anonymous dataset

Rec. #	Birthdate	Gender	Postcode	Disease	Salary
R1	1982–1990	Adult	345**	Gallstones	200,000
R2	1982–1990	Adult	345**	COVID-19	300,000
R3	1982–1990	Adult	345**	COVID-19	400,000
R4	1982–1990	Adult	345**	Stomach cancer	200,000
R5	1982–1990	Adult	345**	Lung cancer	200,000
R6	1982–1990	Adult	345**	COVID-19	200,000
R7	1982–1990	Adult	345**	Malaria	200,000
R8	1982–1990	Adult	345**	Malaria	200,000

attribute, this would also satisfy l-diversity, but also increases the probability of Alice correctly guessing which record belongs to Bob.

A further issue that can arise based on Table 2.13, is the risk of a **similarity attack**. In this case if we further augment Table 2.13 to include a *Salary* (annual salary) as a second sensitive attribute, we get Table 2.14. Suppose, Alice guesses that Bob's record has a high probability of being in equivalence class#1, then she can also infer that Bob's annual salary is the range of 200,000– 400,000 which is relatively high. So Alice can conclude that Bob is a high earner and either has a stomach related condition, or COVID-19.

As Li et al. [123, 124] point out, one of the primary reasons behind both privacy vulnerabilities is due to the fact that l-diverse anonymisations do not take into account closeness of sensitive attributes in terms of semantic similarity and also distributions of values. Li et al. propose addressing this with (n, t)-Closeness [124] and t-closeness [123] which we discuss in the following section.

2.8 t-Closeness

The t-closeness anonymisation principle seeks to ensure that in each equivalence class and Table, the distributions of sensitive attribute-values is such that the degree of similarity is no greater than a pre-defined threshold. If all the equivalence classes in a dataset satisfy t-closeness, the dataset is considered to be t-close anonymous. To measure the distance between distributions t-closeness employs the earth mover's distance. In order to handle categorical or textual attributes a classification hierarchy or taxonomy is used, similar to what is done for standard generalisation mechanisms, with the difference that distances are calculated based on the depth of the tree. Since computing the earth mover's distance can be computationally intensive especially for larger datasets, Li et al. [124] proposed a more relaxed version of t-closeness, namely (n, t)-closeness where by an equivalence class is said to have (n, t)-closeness if it adheres to the following definition.

Definition 2.8 (**(n, t)-closeness [124]**) An equivalence class E_1 obeys (n, t)-closeness iff $\exists$ an equivalence class E_2 such that $E_1 \subset E_2$ where the cardinality of $E_2 = n$ and the distance between the distributions of sensitive attribute-values in E_1 and $E_2 \leq$ a threshold value of t. A table T is (n, t)-close iff all E_i are (n, t)-close.

Example 2.7 For a given table T of size 1000 records, in order to satisfy $(1000, 0.1)$-closeness anonymity, we could split the table into two equivalence classes say E_1 and E_2 each of size 600 and 400 records respectively. E_1 is defined by the quasi-identifier (Postcode: 345**; Birthdate: 1982–1990; and Gender: "Adult") and contains 600 records and the second equivalence class E_2 is defined by the quasi-identifier (Postcode: 345**; Birthdate: 1982–1987; and Gender: "Male" or "Female"). If the distributions of both E_1 and E_2 are each close to (i.e. ≤ 0.1) the distribution of the entire table of 1000 records, then the table is said to satisfy $(1000, 0.1)$-closeness.

One of the drawbacks of both variants of t-closeness anonymity is that they both strive to minimise information loss, which in turn can be exploited to de-anonymise t-closeness and similar variants of anonymised data [235]. We discuss this in some more depth in the following section.

2.9 Minimality Attacks

Wong et al. [232, 233, 235] presented the minimality attack, which takes advantage of the goals of minimising information loss in k-anonymity and its variants such as l-diversity and t-closeness anonymisation algorithms. The minimality attack works as follows.

Suppose we have a Table T that contains: (1) a quasi-identifier (QID); and (2) sensitive attributes. In line with our discussions in the previous Sections, an anonymised table T_a, is obtained by subjecting T to a series of transformations aimed at eliminating personal identifiers, as well as processing both quasi-identifiers and sensitive attribute values to eliminate the risk of re-identification of individual tuples.

An example of how a minimality attack can be provoked appears in Table 2.15, which contains nine (9) tuples in total, with two quasi-identifiers: $q1 = \{[30 - 35]$, Physician, Human$\}$ and $q2 = \{[36 - 45]$, Lawyer, Male$\}$. Table 2.16 provides a version of Table 2.15 in which the explicit identifiers have been removed and the tuples classified into three equivalence classes based on quasi-identifier similarity expressed by $q1$ and $q2$. By l-diversity [139] and/or t-closeness [123], the proportion of sensitive values in the class for $q2$ is $\frac{1}{3}$ and by generalisation we can also obtain a proportion of $\frac{2}{3}$ in the class associated with $q1$.

Now, suppose we have a powerful adversary who knows what method was used to generate the anonymised table T_a and in addition has access to T_a. The adversary might deduce that since generalisation must have already been applied to T_a, this implies that there is at least one equivalence class whose quasi-identifier

Table 2.15 Original table (T): Example of un-anonymised and anonymised table with 3-diversity

Name	QID	Disease
Alice	q1	COVID-19
Bob	q1	Flu
John	q1	Cough
Jeremy	q2	Lung cancer
Andre	q2	Gallstones
James	q2	Malaria
Brad	q2	Thyphoid
Dan	q2	Flu
Dan	q2	Stomach ulcer

Table 2.16 Anonymised table (T_a) with 3-diversity

Name	QID	Disease
xxx	q1	COVID-19
xxx	q1	Flu
xxx	q1	Cough
xxx	q2	Lung cancer
xxx	q2	Gallstones
xxx	q2	Malaria
xxx	q2	Thyphoid
xxx	q2	Flu
xxx	q2	Stomach ulcer

was transformed to satisfy 3-diversity. The adversary can infer information such as—(1) at least one of the tuples associated with $q1$ is either male or female hence the reason for the generalisation of the gender attribute values to "Human"; (2) all physicians in the table might be infected by some variant of COVID-19, based on the publication date of the table. The adversary could then conclude that the class associated to $q1$ either contains one male or one female and that given that all the tuples list "Physician" as a profession, it is quite likely that they are all infected by COVID-19.

Suppose additionally that the adversary has access to some auxiliary data embedded in a table, say T^{ext} as given in Table 2.17 and that T^{ext} is a superset of T_a—Table 2.16. In this case, there is no tuple for Johan, Jonas, and Mary in either Tables 2.15 or 2.16. If the adversary knows that $q3$ cannot be generalised, then he/she can infer that tuples with quasi-identifier $q3 = \{[30-45]$, Teacher, Male$\}$ do not exist in T_a. Therefore, the extra information present in T^{ext} does not affect the inferences drawn based on the data in Table 2.16.

2.9.1 Discussions

However, we also note that Table 2.17 provides information such as that the participants in equivalence class 4 might quite likely have travelled to a tropical

Table 2.17 Example of external table T^{ext} where $T^{ext} \supset T_a$

Name	QID	Disease
xxx	q1	COVID-19
xxx	q1	Flu
xxx	q1	Cough
xxx	q2	Lung cancer
xxx	q2	Gallstones
xxx	q2	Malaria
xxx	q2	Thyphoid
xxx	q2	Flu
xxx	q2	Stomach ulcer
Johan	q3	Thyphoid
Jonas	q3	Malaria
Mary	q3	Lassa fever

country, and that tuple t_6 might also be associated to someone who was in the same travel group but of a different age category.

Minimality attacks are enabled mainly by the anonymisation scheme's goal of minimising information loss in order to maximise data utility. We consider other variants of de-anonymisations that are enabled by the goal of low information loss, in the next section.

2.10 Semantic Attacks and Others

Semantics-based attacks [123, 139, 235] go a step further to exploit meanings deduced both from the data available in the anonymised dataset (foreground knowledge), and from correlations to external sources or datasets (background knowledge). Building on the example we have just given for how minimality attacks occur, an example of a semantic attack emerges when we consider what happens when T^{ext} is a superset of T_a; and $q3 = \{[30-35], \text{Teacher, Male}\}$. For instance, from the 3-anonymous patient dataset in Table 2.17, an adversary conduct a table linking attack by using the information available in Table 2.16 to compute that the probability of Alice's data being present in Table 2.17. Since there are three (3) records in Tables 2.15 and 2.17 containing the quasi-identifier ([30–35], Physician, Human) the likelihood is calculated as $\frac{3}{3} = 1,0$ to indicate that there is a high probability of Alice's record being in equivalence class #1. Similarly for Bob, that the probability of him being present in Table 2.17 but not in the equivalence class associated to $q1$ is $\frac{6}{9} = 0,67$. So, in essence, the adversary can deduce with a high probability of whether or not Alice and Bob are COVID-19 positive based on his/her access to Tables 2.15 and 2.16.

There are a number of other attacks to k-anonymity that are discussed in the literature [45, 123, 139, 161, 212, 232, 233, 235]. Most however are centered around distributions of data values on a per equivalence class and/or general dataset basis. For instance, Wong et al. [231] propose (α, k)-anonymity to protect against re-identifications and relationships to sensitive information in data; and in addition propose global and local recoding to minimise data distortion. This is in a similar vein to solutions such as Incognito [118] which focuses on local recoding, and Mondrian [119] which addresses anonymisation of multi-dimensional (and high dimensional datasets which we discuss in subsequent chapters) datasets. As mentioned before, this is useful from the data utility perspective in terms of supporting privacy preserving data analytics, but attacks such as the minimality attack [233, 235] highlight the fact that privacy exploits can take advantage of this.

2.10.1 δ-Presence and Missing Values

In order to deal with re-identifications in composed datasets, Nergiz et al. [161] propose using a δ-presence metric to quantify the risk of poor anonymisation in datasets. The δ-presence metric or measure evaluates the risk of privacy disclosure based on the distributions of individuals. This is both in equivalence classes but also in the entire dataset. This bears some strong similarities to t-closeness anonymisation but has the added benefit of quantifying privacy risk in terms of the information gain accessible to the adversary. Composition attacks, build on schemes such as Nergiz et al.'s and reverse engineer privacy risk evaluations in order to determine which de-anonymisation goals hold the most reward (e.g. the most de-anonymised data releases) [84]. An added dimension of composition attacks is that the adversary uses as much data as can be accessed and combined, to obtain a de-anonymised version of the target dataset. This is also applied intuitively in random de-anonymisations such as the AOL data de-anonymisation [97].

2.10.2 NULL-Aware Generalisation

Ciglic et al. [45, 46] also present various de-anonymisation attacks that exploit the presence of NULL and/or missing values in relational k-anonymised, their implications on inferences drawn from quasi-identifier attribute values. We have discussed generalisation and suppression as a means of enabling the formation of equivalence classes and by extension, anonymised data. However, one of the caveats that emerges in applying either one or both of these transformation measures is that missing information (due for instance to suppressions, or generalisations to a generic value such as "*") is typically represented in relational database tables with a NULL value.

The k-anonymity approach as described by Samarati and Sweeney [193] works on the premise that tuples containing NULL values are removed from the dataset (microdata). Ciglic et al. refer to this as the "Basic Match" property. For illustration, we modify our running example (see Table 2.2) to obtain Table 2.18 which

Table 2.18 Original dataset with NULL values and augmented quasi-identifier (gender, age, job, postcode)

	Job	Age	Gender	Postcode	Disease
R1	NULL	34	Female	34520	COVID-19
R2	Mayor	32	Male	34521	COVID-19
R3	Doctor	33	Male	34522	COVID-19
R4	NULL	37	Female	34520	Lung cancer
R5	Technician	NULL	Female	34520	Gallstones
R6	Teacher	NULL	Female	34520	Malaria
R7	NULL	42	Male	34520	Thyphoid
R8	NULL	43	Male	34520	Lung cancer
R9	Doctor	44	Male	34525	Stomach cancer

Table 2.19 Original dataset with NULL values and augmented quasi-identifier (gender, age, job, postcode)

	Job	Age	Gender	Postcode	Disease
R2	Professional	[30–40]	Male	3452*	COVID-19
R3	Professional	[30–40]	Male	3452*	COVID-19
R9	Professional	[30–40]	Male	3452*	Stomach cancer

includes NULL values, a quasi-identifier $Q' = Gender, Age, Job, Postcode$, and a sensitive attribute: "Disease".

To obtain a 3-anonymous table, following the Samarati and Sweeney approach [192, 193], we remove all the rows containing the NULL values, and then apply generalisation as well as suppression to obtain Table 2.19

In this case (see Table 2.19) we see that the information loss incurred is very high in that 2/3 of the tuples in the table have been suppressed, furthermore the remaining attribute values have been modified to ensure 3-anonymity based on the quasi-identifier. As mentioned before, k-anonymous datasets such as these are aimed at preventing privacy disclosures due to tuple linking attacks. Background knowledge attacks might also potentially be possible based on whether the adversary has knowledge of whether or not a James (who identifies as male) and lives in the 3452* area could be associated to the tuple R9 Professional, [30–40], Male, 3452*, Stomach Cancer and in the case reconstruct the original values in the tuple James, Doctor, 44, Male, 34525, Stomach Cancer. This could be alleviated with some of the approaches such as l-diversity and t-closeness, which we have discussed.

Suppose that to alleviate information loss, we decided to forego suppressing the NULL values. In this case, we could modify Table 2.18 to obtain a 3-anonymous table—Table 2.20 by applying generalisation and suppression aimed at retaining NULL values, in order we assume that NULL represents a missing value or a negation. For instance, referring back to Table 2.18 the NULL value at R5 for the Age attribute is NULL because the person concerned falls into the [30–40] age category but wishes to conceal the actual value.

Table 2.20 Original dataset with NULL values and augmented quasi-identifier (gender, age, job, postcode)

	Job	Age	Gender	Postcode	Disease
R1	NULL	[30–35]	Prefer-Not-To-Say	34520	COVID-19
R2	Mayor	[30–35]	Prefer-Not-To-Say	34520	COVID-19
R3	Doctor	[30–35]	Prefer-Not-To-Say	34520	COVID-19
R4	NULL	NULL	Female	34520	Lung cancer
R5	Technician	NULL	Female	34520	Gallstones
R6	Teacher	NULL	Female	34520	Malaria
R7	NULL	[40–45]	Male	3452*	Thyphoid
R8	NULL	[40–45]	Male	3452*	Lung cancer
R9	Doctor	[40–45]	Male	3452*	Stomach cancer

Table 2.21 Original dataset with NULL values and augmented quasi-identifier (gender, age, job, postcode)

	Job	Age	Gender	Postcode	Disease
R4	NULL	NULL	Female	34520	Lung cancer

The question then arises as to whether or not NULL values can result in privacy disclosures. To this end, we consider the NULL identifier attack which employs a principle similar to negative disclosure in *l*-diversity, in that the adversary exploits knowledge that a certain value does not exist (e.g. a NULL value). Assume that the adversary knows that Tina is unemployed, is terminally ill, and lives in the 34520 postcode area. The adversary can query table with a search condition such as Gender = "Female" and Job = "NULL" which will give the following result (see Table 2.21):

So including NULL values in the dataset does not guarantee privacy against background knowledge and NULL-identifier attacks. The same is also true when NULLs are treated as missing values [45, 46].

One other attack worth mentioning is the singularity attack which does not depend on the type of NULL values (whether they are missing, prefer-not-to-disclose, or non-applicable) but rather attribute values that are unique by default. For instance, Postcode and Job could be employed for disclosure if Job = "Neurologist" and the adversary knows that every Postcode area has only one "Neurologist".

2.10.3 Discussions

It is worth noting however, that while we have discussed the vulnerability of schemes based on the philosophy of *k*-anonymisation to linking and re-identification attacks, as Cormode et al [49] point out, it is possible to minimise the impact of such attacks by (1) bearing in mind that these schemes can only guarantee sub-optimal privacy since the algorithms have been shown to be NP-hard (see Sect. 2.16); and (2) The confidence about discovering the presence of an individual in a well-anonymised dataset is very low in practice without this negatively impacting utility. However, as we note in the discussion that follows, guaranteeing even sub-optimal privacy is a challenging problem [109].

In the next sections we consider other types of data that are usually composed by profiling users in online scenarios. Such datasets can be represented using the relational model that we have focused on so far, but are also extensible to other data representations (models). We start in the next section by considering movie data, and the special case of Netflix data.[1]

2.11 De-Anonymising Netflix Data

One of the main reasons for publishing microdata is collaborative filtering whereby a user's future choices or decisions are predicted from past behaviour based on user similarity. The main goal is to predict the value of some attributes using a combination of information drawn from other attributes. Application examples include shopping recommender systems like Amazon, aggressive caching in Web browsers, and online education platforms.

Large sparse datasets are generally sparsely populated in that each individual tuple contains only a small fraction of attribute-values. As such, achieving privacy by anonymising such datasets is a challenging problem, because each tuple can have a unique quasi-identifier thereby making cluster (equivalence class) formation difficult. one example of large sparse data is Netflix data.

2.11.1 De-Anonymisation Mechanism

In order to de-anonymise large sparse data, Narayanan and Shmatikov[159] proposed a de-anonymisation algorithm that includes three components namely, a *scoring function*, a *matching criterion*, and *record selection*. The scoring function assigns a numerical score to each tuple in the subset T' of a table T, that is under evaluation for the potential for de-anonymisation. The numerical score indicates how well the tuple in T' matches the adversary's auxiliary information. The matching criterion builds on the scoring function to determine whether there is a match between the scores on the tuples in T' and the data in the adversary's auxiliary information Aux. The record selection mechanism selects a "best-guess tuple" or a probability distribution for the range in which the correct guess might lie.

To determine the feasibility of the de-anonymisation algorithm, the authors [157] used the 2006 Netflix dataset which had already been anonymised and publicly released. The dataset contained 100,480,507 movie ratings created by

---

[1] Netflix is a popular online film service provider, in western countries.

480,189 Netflix subscribers between December 1999 and December 2005. The de-anonymisation algorithm was successful in identifying subscribers in the dataset based on some background knowledge obtained from repositories such as the Internet Movie Database. A further issue was that the algorithm could even go further to uncover deep private information about the subscribers, such as political preferences and religious views.

For instance, using the similarity scoring function, with 8 movie ratings (of which two may be wrong) and dates with an error margin of $\pm$ 14-days, 99% of records can be uniquely identified; and 68% using 2 ratings and dates with an error margin of $\pm$ 3-days. Furthermore, even when knowledge of the ratings is not available up to 84% of subscribers can be identified if the adversary know 6 out of 8 moves that fall outside the top 500.

2.11.2 Discussion

What is worth noting from the Netflix de-anonymisation is that strong anonymisation algorithms and the data generated therefrom need to be supported by robust outlier detection and analysis mechanisms. In other words, the presence of outlier data points in the Netflix dataset enables the de-anonymisations. Therefore using such datasets in a privacy preserving way requires that the data analytics mechanisms be enabled with continuous outlier detection and analysis, to determine what anomalies persist in the dataset and that might be potentially exploitable for de-anonymisations.

However as we discuss in the following section, auxiliary data plays an important role in enabling de-anonymisations and so effective privacy preserving data analytics mechanisms cannot ignore such datasets.

2.12 De-Montjoye De-Anonymisation

De-Montjoye et al. [151] studied three months of credit card data containing 1.1 million tuples and showed that only four spatio-temporal points are needed to uniquely identify 90% of the tuples in the dataset. This corroborates the Sweeney et al. [210] result as well as the Golle result [93] regarding the privacy of pseudonymised data. Furthermore, it is worth noting that even high dimensional datasets that provide coarse-grained information provide little or no anonymity. Gender was also cited as strong proponent for re-identification. For instance, a dataset with a high proportion of females carries a higher de-anonymisation risk than one with males.

Both results emphasise the importance of quasi-identifier generalisability in ensuring privacy, and also that outlierliness as discussed in the Netflix data case, is a strong determinant in de-anonymisations.

2.12.1 De-Anonymisation: Considerations

One of the key metrics used to enable de-anonymisations of individuals in such datasets is *unicity* [151]. *Unicity* provides a method of expressing how much information is needed from auxiliary datasets and/or background information, in order to re-identify an individual.

The schemes we have discussed in the previous sections, work precisely to reduce the probability of re-identification by minimising the *unicity* of individuals within a given anonymised dataset.

2.12.2 Per Use Case Anonymisation: Successes

Furthermore, as Sanchez et al. [195] pointed out, and as we discuss in subsequent chapters, anonymisation can be conducted reliably by well-established techniques in the literature [143, 163]. In each case, the goal is to minimise the degree of individual *unicity*, while minimising the degree of information loss a bid to retain the utility of the data for data analytics operations. This needs to be done on a per case basis, but also bearing in mind the potential use-cases of the data.

We present some solutions in Chaps. 3–8 which leverage heuristic to enable effective data anonymisations on a per context and per privacy preserving data analytics use case.

2.13 De-Anonymising Web Browsing Data

So far we have discussed privacy from the perspectives of datasets in terms of how they are collected and what enables de-anonymisations, as well as the considerations needed to support users in configuring privacy settings on applications in order to support collecting privacy preserving data. However, since data is a driver for most if not all services on the Internet, it is also important to consider whether or not it is possible for online trackers and network adversaries to de-anonymise web browsing data for instance. This relates back to the de-anonymisation incident with AOL search data [22, 97] and also to the Netflix dataset de-anonymisation (see Sect. 2.11) [157] and the De-Montjoye de-anonymistion.

Linkage attacks based on user behaviours are important to consider in terms of generating anonymised datasets because they impact on several data types that are collected daily on Internet applications. Examples include transaction data [157], location data [137, 150], credit card metadata [151], and textual data (writing/communication style) [159].

Several works have been conducted to study the general problem of third party tracking online [38, 128]. For instance, results from a study that Libert

[128] conducted, indicate that Google can track nearly 80% of websites on the Internet through third-party domains. Yet, in spite of legislation such as GDPR, web tracking has broadened its scope beyond HTTP cookies to include persistent tracking mechanisms such as E-Tage [5], HTML5 localStorage [5], Cookie syncing [21, 72], Nudging [59, 183], and device fingerprinting [6, 71, 72].

Some of the initial works on tracking and how web browsing data can be employed to support de-anonymisation include Olejnik et al.[165] work. Olejnik et al. conducted a large online experiment involving 370,000 users to demonstrate that behavioural fingerprints on the web stabilise over time and so, can serve to support de-anonymisations by browser history sniffing.This corroborates the idea that large sparse datasets increase unicity and outlierliness thereby making de-anonymisations easier. The results showed that testing 50 links is sufficient to uniquely identify 42% of the users in the experiment sample. However, Olejnik et al.'s results do not indicate whether or not such behavioural fingerprints can be linked to auxiliary data, such as on social media, to support de-anonymisations. Wondracek et al. [229], build on history sniffing to leverage vulnerabilities on social media platforms such as user comments and likes, to show that group memberships on social media tend to be unique. However, given that most of these vulnerabilities have been fixed on existing social media platforms, the risk of such de-anonymisations is significantly reduced.

An important aspect to highlight is the fact that online trackers and network adversaries can de-anonymise web browsing data by associations between browsing history and social media profile data, is that conducted by Su et al. [206]. For instance, using theoretical simulations and experiments on real user data, Su et al. [206] showed that de-identified web browsing histories can be linked to social media profiles using only publicly available data. The basis of the adversarial model is that most users on social media, tend to click on links posted by accounts that they follow or that belong to users in their friendship network, and these click patterns persist in the users' browsing histories. This work also emphasises how leakages of personal identifying information can further foil the goals of pseudonymisation schemes in providing privacy (see Sect. 2.3) [72].

We now discuss de-anonymisations in social network graphs, to show case examples of how information sharing platforms enable de-anonymisations through auxiliary data in the form os personal information.

2.14 De-anonymisations of Social Network Graphs

Social media platforms provide a popular source for user data that lends itself well to several purposes such as market analysis, research, and statistical analysis [239]. One primary reason for this, is that users (data owners) tend to share very rich sources of personal data ranging from photographs to even telephone numbers shared in chat messages and/or posts [122]. Adversarial parties wanting to foil security mechanisms therefore access social media platforms to look for clues to enable password guessing attacks and impersonations to enable identity theft, for instance [60, 81, 86, 138, 158, 180].

Several works have proposed ways, primarily centered around clustering, graph modification, and differential privacy to anonymise social media graphs [1, 19, 42, 85]. As is the case with the anonymisation approaches we have discussed so far, the goal of these techniques is to preserve data usability to facilitate data analysis while concealing the identities of the data owners (privacy). We now show that these approaches are also vulnerable to the issues we have discussed so far, in a bit to highlight the fact that the privacy exploits we have discussed are translatable to other kinds of data such as graph data.

The embedded subgraph attack [23] allows the adversarial party to identify whether relationships between users exist in an anonymised social network. To conduct the attack the adversary selects a set of users say, John, Alice and Bob, whose privacy can be violated either due to mis-configurations of privacy settings or inadvertent disclosures of private personal information. The adversary enables the attack by creating a small set of sybil nodes and uses these nodes (social network user profiles) to contact the target users.

Considering the case of Alice, Bob and John, and that an adversary say Dick wishes to de-anonymise the friendship graph that Alice has with Bob and John. Dick then proceeds to create a sybil network with fake or real users Tom and Harry. Next, Dick then randomly contacts Alice, Bob, or John requesting a friendship. If Dick succeeds he then proceeds to embed the malicious subgraph (Tom and Harry) into the social network graph of Alice, Bob, and John. In this way, Dick can freely access content posted by Alice, Bob, and John.

The embedded subgraph attack can be extended to enable attacks such as the coalition attack, in which the adversary exploits small uniquely identifiable subgraphs. Revisiting the example with Dick, Dick could collude with Tom and Harry to form a fictional social network graph. Tom, Dick, and Harry could then contact other users to form friendship graphs which by transitive disclosure would reveal information from those friendship graphs to Tom, Dick and Harry, even when they are not directly connected to the said friendship graph. In this was the adversary, say Dick, gains knowledge about the friendship graphs associated both to himself and Tom and Harry, by proximity and inference, enabled by transitive disclosures. In a similar vein to how auxiliary information enables de-anonymisations, Dick could also collect information about a target user say, Bob by collecting information about the friends (people) who might potentially be in Bob's friendship graph [158].

The main issue that arises with all de-anonymisations of social network graphs is that the attacks are enabled by the fact that the adversarial party can use the rich source of information presented on social media platforms to infer information about a target user. As such it makes sense, and this we do in the next section, to discuss the concept of trusted and untrusted data collectors and the implications for enabling the generation of reliable privacy preserving data.

2.15 Trusted vs. Untrusted Data Collectors

Having discussed methods of creating privacy preserving data, it is important to wrap up this chapter with a note that a lot of research has been conducted on the subject of data privacy in both the computer science (data mining, machine learning, data science,...) and statistics communities. A lot of the work can generally be classified under two main categories namely: *Trusted Data Collector* and *Untrusted Data Collector*.

2.15.1 Trusted Data Collector

In the trusted data collector (also sometimes referred to as the "Honest-but-Curious" data collector) scenario, users (data owners) trust the data collector not to breach their privacy. Examples of trusted data collectors include government agencies, hospitals, banks, and education institutions. In some cases, however, these agencies might need to share the data with third party service providers. For example, a hospital might need to share a patient's healthcare data with a health insurance provider. Ensuring that the sharing operation provides the required information without breaching the privacy of the individual involved is an important consideration. Several methods of publishing such datasets have also been widely discussed in the literature. We focus on four classes namely: (1) Publicly available microdata; (2) Data on response to allowed queries; (3) Shared Data with Authorised Parties; and (4) Searching over encrypted data.

2.15.1.1 Publishing Microdata

Research on publishing microdata has focused on providing data utility and guaranteeing privacy. Work in the statistics community that is centered around census data has focused on identifying and sensitive data protection in contingency tables representing a cross-classification of the data [139, 143]. Some of the main approaches that have been employed to protect sensitive data in this case include: *data swapping*, *data suppression* and differential privacy. Data swapping entails moving data entries from one cell to another while maintaining an overall consistency of the data [222]. Data suppression by contrast, simply deletes cells with low counts of data [117]. Since suppression can lead to high information loss and swapping, can create high distorted data which in turn impacts negatively on the utility of the data, some of the approaches that have been proposed to alleviate this issue employ safety ranges or protection intervals on a per attribute-value or attribute basis [34, 117]. Other approaches such as differential privacy employ perturbation to distort the data with statistical noise additions.We have also discussed in previous sections some of the approaches based primarily on syntactic and semantic data anonymisations that can be employed to transform microdata for publishing publicly [34].

Other works like that of Chawla et al. [43] proposed achieving privacy by "blending in a crowd", which operates on a principle similar to clustering as we have discussed in k-anonymity. In this case, the adversary's job in terms of isolating tuples is made difficult by employing a similarity metric that enforces indistingusihability. Similarity is computed by means of perturbation and histogram techniques for data sanitization. However, because the scheme is heavily reliant on distance measures for privacy, it fails to identify scenarios in which single data points might result in re-identifications. This is similar to the issue of de-anonymisations discussed in relation to sensitive data distributions (see Sect. 2.6). Miklau and Suciu [146], as well as Yang and Li [240], also present other approaches centered on characterising views that are publishable from a dataset, which we also discuss in some more depth in Chap. 4, and methods of matching access control authorisations to privacy. In all of these cases, several issues arise—first, achieving optimal privacy from the theoretical perspective can sometimes be too strict and result in high information loss; and second, how to deal with the possibility that the adversarial party might have background knowledge about the published data and so could potentially successfully breach privacy.

2.15.1.2 Data on Response to Allowed Queries

The trusted data collector can operate as a query answering service. In this case, the database answers primarily aggregate queries, such as `COUNT`, `SUM`, and `AVG` for a specified subset of tuples (i.e. assuming a relational database model). The goal is to provide a neutral query response to avoid either positive or negative disclosures [123, 124, 126, 139]. In order to achieve this, techniques for statistical database query answering are broadly classified into three (3) categories namely: (1) *Query Restriction*, (2) *Query Auditing*, and (3) *data and output perturbation*.

Query restriction schemes operate by restricting the types of queries that should not be answered in order to ensure that privacy is not breached. Responses to legal or whitelisted queries are not perturbed. In this case aggregation functions are used to respond to queries on a per usecase basis [65]. One key caveat of this scheme is that historical tracking mechanisms [57], can be used to learn values of sensitive attributes. This can be remedied by using query set overlap techniques can be used to disallow queries that have a large overlap with previous queries. However, this remains a challenging problem especially in the light of complex datasets such as data meshes and high dimensional data which appear increasingly in web scenarios [177].

Query auditing involves determining whether or not to respond to a query based on what queries have been submitted to the database beforehand. For example, the database might respond to query (q_i) if and only if the responses to previous queries $(q_0, \ldots, q_{i-1})$ do not conflict with the response to q_i [65, 68]. This process can be conducted online or offline, but raises advantages and disadvantages from the performance efficiency perspective that must be evaluated on a per processing scenario basis [106]. One advantage is that by historically monitoring releases we can control the risk of de-anonymisation. However, growth in query numbers and dataset sizes can potentially become a performance bottleneck.

Finally data perturbation techniques answer queries on the basis of a perturbed version of the data. However, responding to queries on the basis of perturbed data raises an issue of bias, in the sense that the query response based on the perturbed data can differ considerably from what it would have been on the original data, which is something to avoid [64, 139]. Other approaches seek to alleviate this issue by running the query over the original data but distorting the responses with noise additions [68]. This is also the primary basis of differential privacy as we have mentioned before [64]. Our work on Base Rate Neglect [174] and on inference discovery [175], but that falls outside the scope of this monograph, also brings this issue to the fore.

2.15.1.3 Sharing with Authorised Parties

The idea of sharing data with authorised parties started off with the notion of Hippocratic databases [15]. Hippocratic databases require that both data owners and data collectors specify the privacy policies that describe the use cases of the data as well the recipients who are authorised to access the data. The policies are specified using a policy specification language [15, 32]. Other considerations taken into account in forming Hippocratic databases are support mechanisms for audit trails, logging, disclosure limitations, and data retention. Preventing and detecting adversarial manipulations to log data is therefore an important problem in relation to maintaining such Hippocratic databases.

2.15.1.4 Searching Over Encrypted Data

Data owners often outsource their data to cloud service providers in order to reduce storage and infrastructure costs, for instance [141]. However, security and privacy are two issues that discourage data owners from adopting this solution. One approach to alleviating this concern is to employ searchable encryption as a means of assuring privacy and confidentiality of such data [114, 222]. The principle behind searchable encryption is to yield a result based on a query but the provided result must not reveal anything other than the search outcome and pattern [20]. In essence, the goal is to provide a query result that adheres to the properties of *Confidentiality*, *Integrity*, and *Availability*. Therefore, a user without the "correct" authorisations should not have access to query results that do not match his/her credentials.

This goes in line with research on problems such a private collaborative computation and secure multi-party computation [68]. Existing work formulated the problem around expressing function as a Boolean circuit, but were critiqued for being inefficient in terms of communication. More recent work has considered aspects such as the commodity server model [65], searching over encrypted cloud data [127], Homomorphic encryption [73], association rule mining and privacy-aware character pattern matching [141] and private clustering [50, 141]. A key drawback in relation to how privacy preserving data analytics is enabled is centered around how searchable encrypted data shares the information disclosed, what sensitive information is output in response to the queries, and what the long term implications are for the privacy assertions that are made on the data.

2.15.2 Untrusted Data Collector

The untrusted data collector composes data based on individual repositories shared by data owners [50, 90]. The data owners randomise their data before it is shared with the data collector. This is in a bid to avoid attacks such as unsorted matching which we discussed in Sect. 2.5.4.3. For example, data collected from individual phones or wearables is randomised using differential privacy mechanisms to guarantee users privacy on Apple eco-systems [50, 101]. Randomisation schemes like differential privacy where correlations in noise additions to query responses match correlations in the data have been shown to protect against de-anonymisations of sensitive data [65]. However, in large data scenarios that are processing intensive and where there is a strong requirement for real-time responses, running queries on demand can be time consuming.

2.16 On NP-Hardness and Other Matters

Anonymisation algorithms are generally performance intensive because of the operations that are needed to transform the data for privacy. For example, applying generalisation and suppression transformation measures to high-dimensional data provokes high levels information loss, rendering the data useless for data analytics. Having methods of efficiently detecting privacy vulnerabilities due to linking (inference) in anonymised datasets and transforming the datasets to prevent de-anonymisations, can support service providers in generating, sharing, and conducting data analytics in a privacy-preserving manner.

Meyerson and Williams [144] as well as Bayardo et al. [27], have shown that both optimal k-anonymisation and suppression (where the goal is to minimise the number of attribute values deleted from a relation) is NP-Hard, but that an $O(k \log m)$-approximation, where m is the degree of the relation can be employed time-efficiently. Xiao et al. [237] have also shown that optimal l-diverse generalisation is NP-Hard even when there are only three (3) distinct sensitive attribute-values in the microdata, and that an (l, d)-approximation algorithm where d is the dimensionality of the underlying dataset, efficiently minimises information loss. Dondi et al. [62]

also show additionally that the l-diversity problem is APX-hard even if $l = 4$ and the input dataset consists of only three attributes namely, a sensitive attribute and two quasi-identifier attributes. Further work by Dondi et al. [62] has also studied the parametrised complexity of the l-diversity problem and shown that the problem is $W[1]$-hard when parametrised by the cost-bound (threshold of information loss and/or privacy), by l (fraction of distinct sensitive attribute values in a given equivalence class), and by the size of the alphabet (attribute values). Liang and Yuan [126] also have demonstrated that optimal t-closeness generalisation for a given dataset is NP-hard for every constant t such that $0 \leq t < 1$.

This supports assertions about sub-optimal privacy and the arguments that promote using heuristics as a means of achieving better approximations to perfect privacy [26, 171, 172].

2.16.1 *Other Privacy Issues*

However, sub-optimal privacy raises the question of how to protect Personal Identifying Information (PII) from exposure in the case of composed data. As we have mentioned before, most de-anonymisations are possible due to unicity that emerges when datasets are combined. This can include ones that are anonymised or pseudonymised with ones that are not [150, 151, 212]. Furthermore legislation such as GDPR, HIPAA or CCPA, requires firm guarantees of protection of personally identifiable information (PII), deletion upon request, and security even in distributed settings. Additionally, generating privacy preserving data compositions requires that we give consideration to the impact that dynamically changing data can have on the privacy assertions that are made about a given combination of datasets and the disclosure risks that emerge therefrom.

Furthermore, while quasi-identifiers might not pose a privacy risk as independent occurrences in a given dataset, combining datasets might result in unintended privacy disclosures. For instance, suppose three different data owners A, B, and C decide to combine their datasets to enable a joint data analytics operation. The data belonging to A contains a quasi-identifier Postcode, while B contains Age, Gender, and C contains Profession, Age. Combining all three datasets essentially creates a new quasi-identifier Postcode, Age, Gender, Profession which could potentially be linked to a dataset D that contains attributes such as Heart Health, Medical Insurance. So it might be possible to build a profile of a given user's health conditions by analysing the data in search of outliers [177].

2.17 Anonymity for Data Mining

In terms of privacy preserving data analytics most existing solutions to deal with the risk of re-identification (de-anonymisation) have considered data transformation for anonymity in the field of machine learning. Existing solutions include regression models [78], where the goal is to explore causal relationships between attribute values and reason about how inferences that enable re-identification attacks occur. Anonymisation schemes such as clustering [220], and naive bayes classification [221] employ unsupervised learning and statistical methods as a means of learning about and/or understanding underlying patterns within the data. The goal is both to support data analytics operations but also to determine what risks of privacy violations appear in the dataset under consideration [150, 151, 156, 206, 223, 243].

In general anonymisation approaches aimed at supporting privacy preserving data machine learning (analytics) can be broadly categorised as being either *syntactic* or *semantic*.

In *syntactic* approaches, the idea is to create a generic dataset that is anonymised to suit a wide variety of needs. In *semantic* approaches, the dataset is kept in its original form. Only when a specific query is submitted to the dataset is an anonymised subset of the original dataset returned in response to the query. *Semantic* approaches are challenging to use in scenarios where the use cases of the data are not known before the queries are issued. So, personal identifying information can inadvertently be revealed. In addition, semantic approaches like differential privacy do not necessarily guarantee anonymity in case of colluding users [109]. Some of the issues with semantic approaches, have been addressed from a practical point of view by Dankar and El Enam [53, 54] in the context of health data, but translating these solutions to other types of large high-dimensional datasets still poses the issue of inference due to quasi-identifiers.

Syntactic mechanisms address the issues of collusion, and inadvertent exposure of personal information by anonymising the datasets to remove as much personal information as possible (privacy preservation), while minimising information loss. The premise here is to use the size of datasets to make de-anonymisation hard by making it harder to distinguish individual records. At the same time the volume of anonymised data available reduces the motivation for collusion. Furthermore, as we have mentioned before, high information loss impacts negatively on data utility [227]. However, existing studies of *syntactic* approaches to anonymisation [65, 123, 139, 213, 231, 232, 232, 233, 235], raise the fact that vulnerabilities to linking attacks such as minimality and semantic attacks that exploit inferences drawn from attribute-value combinations, make generating low information loss privacy preserving datasets a challenging problem.

Two of the primary inference attacks that arise due to low information loss in syntactically (a-priori) anonymised datasets include ones based on *Minimality* and *Semantics*. Attacks based on the notion of *minimality* exploit the goal of a-priori anonymisation mechanisms of minimising information loss, to infer hidden information based only on the information available in the anonymised dataset. While attacks that exploit *semantics*, go a step-further to infer hidden sensitive information by exploiting the meanings of the values in the anonymised data and by making associations with other values either in the anonymised dataset, or in auxiliary data [24, 45, 178, 184, 231].

The next chapter presents an approach to anonymising data in low powered and low processing resource settings. We build on the assertions of Sanchez et al. [194, 195] to show that sub-optimal anonymised data can be obtained and used safely in a privacy preserving way for data analytics operations.

Part III
Anonymisation Approaches

Chapter 3
Multi-Objective Anonymisation

Oftentimes in designing privacy preserving algorithms, the primary assumption is that collected datasets will be analysed by the data owners or by data collectors who have explicitly obtained permissions from the data owners to do so on the condition that privacy criteria are adhered to. In this chapter we consider a data owner scenario in which a data owner wishes to anonymise his/her data before it is shared with a third-party analytics service provider and/or data collector. In such scenarios, the data owners might not have the on-device (or in-house) processing power to conduct the data analytics operations. As such, the data analytics and storage tasks must be outsourced to a third-party data service provider with the resources to store and/or provide useful insights from the data. To enable data analytics, and in line with the theme of this monograph, a necessary pre-processing step is to anonymise the data before it is shared. However, since the data anonymisation operation is conducted on a comparatively low-powered and low-processing device, implementations of anonymisation algorithms for these contexts need to be refactored accordingly. This Chapter presents an approach for doing so [107, 108].

3.1 Low Powered/Processing: Considerations

One key challenge that appears in low powered/processing contexts is that of effectively, and efficiently analysing large volumes of data. For instance, the growing use of smartphone applications to collect health related data raises the question of privacy. Discussions centered around what and how much personal information users inadvertently release to third-party service providers, remains a concern. Offering users some control over what data gets transmitted and in what form, can offer users a means of enforcing user-end privacy. In many current solutions, service providers opt not to store any personally identifying information.

A. V. D. M. Kayem, *Enabling Privacy Preserving Data Analytics*, Advances in Information Security 92, https://doi.org/10.1007/978-3-031-93906-8_3

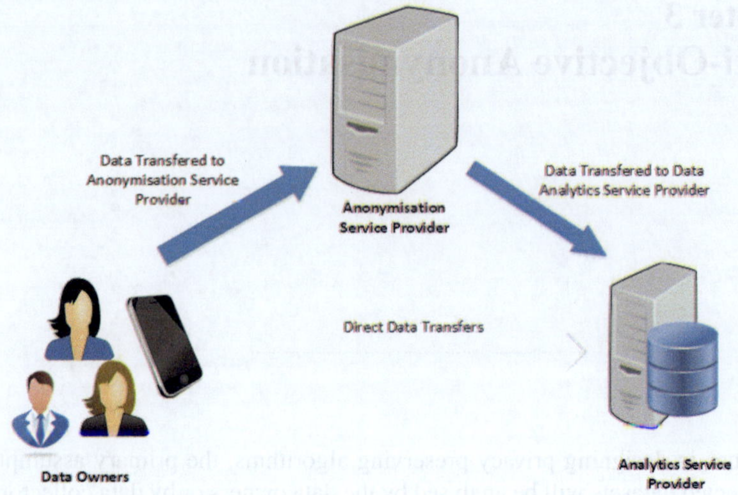

Fig. 3.1 Outsourcing anonymisation architecture

However, as we discussed in Chap. 2, simply excluding personally identifying information does preclude de-anonymisation.

Anonymising the data before it is outsourced to a third party data analytics service provider a cost effective management solution to this problem [39]. An outsourcing architecture to support data analytics in such a low powered/processing scenario could be designed as illustrated in Fig. 3.1.

As shown in Fig. 3.1, we assume the presence of low powered/processing devices such as general IoT (Internet-of-Things) devices and mobile/smartphones that collected and process data based on the per application/device settings activated by a user. The data is anonymised on the device before it is transferred to a third-party anonymisation service provider where it is combined with other datasets submitted by various other data owners. The anonymisation service provider cross-checks the existing data submissions for disclosure risks and shares the data with a data analytics service provider.

One method of enabling on-device anonmyisation is to employ an automated anonymisation scheme that extends the standard k-anonymisation and l-diversity algorithms by using multi-objective optimisation to maximise information utility (minimise information loss) [107]. Disclosure risks, in terms of privacy violations due to linking and inference, can be evaluated with a weighting metric based on information exposure severity. multi-objective anonymisation (MOA) combines techniques from k-anonymity inspired algorithms and works by composing clusters of size k based on quasi-identifier similarity. In order to prevent de-anonymisations of data, multi-objective anonymisation selects the cluster formations that provide the lowest de-anonymisation risk. To do so, a severity ranking (cost to privacy due to attribute exposure) is used to evaluate privacy exposure risk and a transformation metric is used to evaluate the cost in terms of information loss due to data distortion. We provide more details on the operation of the MOA scheme in the following section.

3.2 Multi-Objective Data Anonymisation

The multi-objective anonymisation (MOA) combines the principles of k-anonymisation, l-diversity, and t-closeness to generate privacy preserving datasets [107, 108]. To determine which is the best effort option to use for a given dataset and anonymisation case, MOA operates by weighting sensitive attributes with respect to the risk of de-anonymisation of individual tuples in a dataset. A further consideration is the determine the optimal size of a cluster (equivalence class) to ensure minimal information loss and maximal privacy. For clarity, we begin by presenting some preliminary definitions and explanations.

3.2.1 Preliminaries

The MOA scheme comprises three (3) algorithms namely: CG-Kanon (a variant of k-anonymity), CG-Diverse (a variant of l-diversity), and t-clustering (a variant of t-clustering. As mentioned before, we assume a relational database representation where a table $T(A_1, A_2, \ldots, A_n)$ represents a dataset (or a subset of a dataset) with a finite number of tuples. The attribute space A of T (columns of T) is such that $\{A_1, A_2, \ldots, A_n\}$ where $a \in A$ represents a specific attribute value in A. A tuple in T is denoted by t such that $t\{v_1, v_2, \ldots, v_n\}$ denotes a projection of the sequence of values in t. In this context we re-define a quasi-identifier slightly to match the notation in this Chapter, and formally express this as follows:

Definition 3.1 (Quasi-identifier) Given a table $T(A_1, A_2, \ldots, A_n)$, a quasi-identifier Q_T of D is a set of attributes $\{A_i, \ldots, A_j\} \subset \{A_1, A_2, \ldots, A_n\}$ such that $\forall t \in T$, $t\{v_i, \ldots, v_j\} \subset t\{v_1, \ldots v_n\}$, $Q_T\{t\}$ is a set of attributes whose values can be mapped to at least one $t \in T$.

This definition describes quasi-identifiers as attributes which independently or when combined can be used to uniquely identify an individual or entity. As mentioned in Chap. 2, most often such identifications are done by combining external data with the published dataset. To prevent such de-identifications, as a first step, we employ the notion of k-anonymity to classify tuples. For our purposes, and also to help with clarity for the discussion on the MOA scheme, we define k-anonymity as follows:

Definition 3.2 (k-anonymity) Let DT be a dataset such that $DT(A_1, \ldots, A_n) \subset T(A_1, A_2, \ldots, A_n)$, with a quasi-identifier $Q_T\{t\}$ associated to DT. DT is said to satisfy k-anonymity if and only if each $Q_T\{t\}$ appears in at least k tuples in DT, where the size of DT is bounded by k.

Based on our definition for k-anonymity, we can provide a corresponding definition for an equivalence as follows:

Definition 3.3 (Equivalence Class) An equivalence class EC, is a set of tuples $\{t_i, \ldots, t_j\}$ such that $\forall\{t_i, t_j\}$, $Q_T(t_i) = \ldots = Q_T(t_j)$.

Finally, *disclosure* is defined as the unintended loss or reduction of privacy for an individual or entity resulting from a publication of anonymised data. In [123], Li et al. note that two types of information disclosure can be identified namely, identity disclosure and attribute disclosure. Identity disclosure occurs when an individual is linked to a particular record in the released table. Whereas, attribute disclosure occurs when new information about some individual is revealed. The released data then makes it possible to infer the characteristics of an individual more accurately than it would be possible before the data release. On the other hand, a sensitive attribute, is an attribute which contains a value that when combined with a quasi-identifier, could result in disclosure. Example 3.1, provides some clarifications of these concepts.

Example 3.1 2-Anonymity Dataset As shown in Fig. 3.2, the equivalence class adheres to 5-anonymity ($k = 5$), and has a quasi-identifier Q_T = {Age, Gender, Postcode, Disease}. For the tuples contained in the equivalence class, the values that comprise the quasi-identifier appear at least five times. Note that in this case that the "Disease" attribute is sensitive.

However, on examining the equivalence class in Fig. 3.2 more closely, it is possible to provoke linking attacks on the data either by inference or by combining

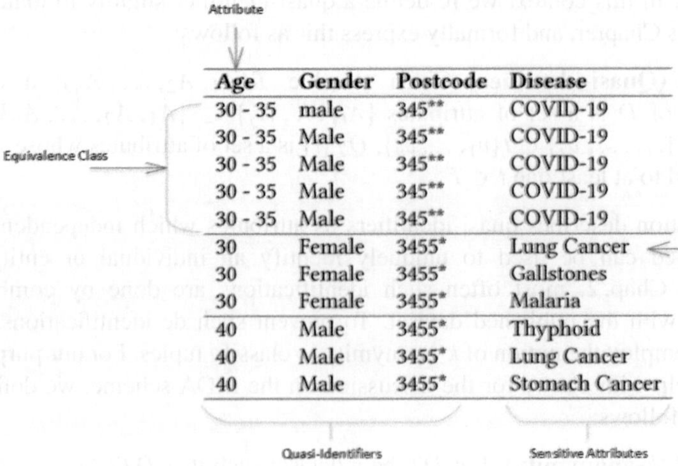

Age	Gender	Postcode	Disease
30 - 35	male	345**	COVID-19
30 - 35	Male	345**	COVID-19
30 - 35	Male	345**	COVID-19
30 - 35	Male	345**	COVID-19
30 - 35	Male	345**	COVID-19
30	Female	3455*	Lung Cancer
30	Female	3455*	Gallstones
30	Female	3455*	Malaria
40	Male	3455*	Thyphoid
40	Male	3455*	Lung Cancer
40	Male	3455*	Stomach Cancer

Fig. 3.2 k-anonymisation: Example

the anonymised dataset with available data from other sources. In Chap. 2, we saw some examples of de-anonymisations that can arise due to k-anonymised data and so we apply l-diversity techniques [139] to alleviate the disclosure risk.

In line with the principles of l-diversity which basically requires that the most frequently occurring sensitive attribute in any equivalence class should not occur more frequently than $\frac{1}{l}$. If we consider the equivalence class in Fig. 3.2 with 5 tuples with "COVID-19" as the most common disease, to achieve 2-diversity "COVID-19" cannot occur more than twice in this equivalence class. Note that another interpretation of this is that there should be at least 2 distinct sensitive values per equivalence class, since any given sensitive value cannot appear more than twice. This definition is referred to as *distinct l-diversity*. Two other stronger forms of diversity, include *entropy l-diversity* and *recursive l-diversity*. *Entropy l-diversity* is based on information entropy, but Machanavajjhala et al. [139] show that this can be too restrictive as it requires a minimum aggregate level of entropy across the whole dataset. *Recursive l-diversity* on the other hand ensures that the most frequent sensitive value occurs frequently enough but the most infrequent value not too infrequently. Since, both of these versions of l-diversity are computationally expensive (in terms of searching through the entire space of sensitive attributes to find suitable SA distributions or combinations), we used a heuristic to categorise sensitive attributes in terms of exposure risk.

CG-Kanon, CD-Diverse, and t-clustering algorithms use generalisation and suppression to support the data transformation process. We now present a modified generalisation and suppression scheme aimed at enabling efficient data transformation on low-powered devices.

3.2.2 Generalisation and Suppression

Let $G(a)$ denote the generalisation tree for numerical attributes and $C(a)$ be the generalisation tree for categorical attributes. Furthermore, $G(a)_{max}$ and $G(a)_{min}$ denote the upper and lower limits respectively for numerical attribute generalisations. While $Max_j^{t_i}(a)$ and $Min_j^{t_i}(a)$ represent the upper and lower limits (leaf nodes) of the generalisation of an attribute a in tuple t_i during the jth iteration of the MOA algorithm. This is because each anonymised dataset obtained is evaluated to determine the cost-benefit tradeoff between information loss and privacy, and we only select the version that offers the best tradeoff between the two. Only when a suitable dataset is obtained, is the anonymised dataset transferred to the third-party service provider.

$C(a)_{total}$ is the total number of leaf nodes generated for $C(a)$ and P is the number of nodes created by $C(a)$. The subtree $C(a)$ of $C(a)$ is rooted at a node $p \in P$ and $c(a)_{p,total}$ is the number of leaf nodes in $c(a)_p$.

In anonymising data, determining a generalisation hierarchy for the quasi-identifiers (Q_T) is a supporting measure in ensuring that the anonymisation

[htbp] for tree= font=, line width=1pt, draw, child anchor=north, parent anchor=south, grow=south, align=center, edge path= [line width=1pt, edge] (!u.parent anchor) —- ([yshift=4pt].child anchor) – (.child anchor) edge label; , , s sep+=20pt, [Taxonomy, fill=white,double=white, [Adult,double=white [Male] [Female]] [Disease,double=white [COVID-19] [Cancer] [Flu]] [Age,double=white [30 - 65 [30 - 45 [30 - 35] [36 - 45]] [46 - 65 [46 - 55] [56 - 65] [NULL]]]] [Postcode, double=white [34* [342*] [345*]] [356* [3567*] [3568*]]]] Generalisation Hierarchy: Disease, Age, and Postcode Taxonomies

Fig. 3.3 Generalisation hierarchy: Disease, age, and postcode taxonomies

	Age	Gender	Postcode	Disease
	Original Data			
	30	Male	34520	Flu
	30	Male	34500	COVID-19
	31	Male	34540	Lung Cancer
	30	Male	34544	COVID-19
	34	Male	34545	COVID-19
	34	Male	34545	COVID-19
	Global Recoding			
Equivalence Class 1	30 - 35	Male	345**	Flu
	30 - 35	Male	345**	COVID-19
Equivalence Class 2	30 - 35	Male	3454*	Lung Cancer
	30 - 35	Male	3454*	COVID-19
	30 - 35	Male	3454*	COVID-19
	30 - 35	Male	3454*	COVID-19
	Local Recoding			
Equivalence Class 1	30	Male	345**	Flu
	30	Male	345**	COVID-19
Equivalence Class 2	30 - 35	Male	345**	COVID-19
	30 - 35	Male	345**	Lung Cancer
Equivalence Class 3	34	Male	34545	COVID-19
	34	Male	34545	COVID-19

Fig. 3.4 Local versus global recoding: Example

algorithm achieves a high classification accuracy whilst minimising the amount of information loss.

As shown in Fig. 3.3 in the generalisation hierarchy, the leaf nodes represent the attribute values and the parent nodes possible generalisations. In Fig. 3.4, for instance by applying generalisation we transform the "Age" attribute value in Equivalence class 1 to 30−35. NULL values are handled without any pre-processing by including a node with a NULL value at each level of the hierarchy [45].

Generalisation is usually accompanied by suppression of portions or whole parts of attribute values, or whole rows in order to prevent linking attacks. For instance, the explicit identifiers like "names" and "telephone numbers" are suppressed from the data. In suppression the previous attribute value is replaced by "*". For instance *firstname* = *"John"* is replaced by *firstname* = "*"; likewise to alter portions of attribute values, and *Postcode* = *"34520"* could be transformed to *Postcode* = *"345**"*.

Attribute generalisation can be done either by using a *global/full domain* generalisation model, or a *local recoding* generalisation model. In global generalisation[118], the occurrence of a specific attribute value is always generalised to the same level. For example, if age 18 is generalised to 18–87 for one tuple, then all tuples with a value of 30 for the age attribute are generalised to 30–35. In local recoding models[212] generalisations are contextualised to equivalence classes. For example, attribute age 30 might be recoded to 30–35 in one equivalence class and to 30–32 in another equivalence class. Once the data has been processed and generalised, the next step is to find a suitable balance between information loss and privacy. Minimising information loss is useful in ensuring data usability while maximising privacy ensures adequate data protection from adversarial access. Since high levels of information loss impact negatively on data utility, a method of minimising information loss in the CG-Kanon, CG-Diverse, and t-clustering algorithms is needed. An information loss model is presented in the following Section.

3.2.3 Information Loss and Severity Weighting

Once the pre-processing is complete, a tradeoff between information loss and privacy must be sought. Anonymised datasets with low information loss are beneficial to data analytics in that the results are close to the ones from the original data. To achieve this goal, MOA employs a piecewise function to minimise information loss on both categorical and numerical data. This is done using Equations 3.1 and 3.2, as follows:

$$IL_i^{t_i}(a) = \begin{cases} \dfrac{k(a)_{p,total} - 1}{P-1} & \text{if categorical} \\ \\ \dfrac{Max_j^{t_i}(a) - Min_j^{t_i}(a)}{G(a)_{max} - G(a)_{min}} & \text{if numerical} \end{cases} \tag{3.1}$$

where the Information Loss Metric is given by:

$$LM^j(a) = \sum_{t_i \in T} \sum_{a \in A} IL_j^{t_i}(a) \tag{3.2}$$

As mentioned before, the proposed MOA scheme aims to support low powered and low processing anonymisation, by allowing end-users to prioritise which attributes or quasi-identifiers should be allowed to retain a high degree of information. We provide a further discussion on other mechanisms for incorporating user privacy preferences, in Sect. 3.5. For now we focus on the MOA heuristic for balancing information loss and privacy.

The MOA scheme, employs a utility function to determine the optimal level of privacy with respect to information loss. The utility function given in Eq. (3.2) employs a weighted information loss metric that is computed through an iterative function until a global optimum is found. At the jth iteration, the weighted information loss metric $LM_{CG,j}$ for the CG-Anon algorithm (that is different attribute combinations for Q_T are evaluated to determine which ones result in the lowest information loss and highest privacy for a given combination of Quasi-Identifiers) is given as follows:

$$LM_{CG,j} = \sum_{t_i \in T} \sum_{a \in A} w_a \times IL_j^{t_i}(a) = w_a \times LM^i(a) \tag{3.3}$$

where w_a is the weight assigned to attribute $a \in A$ by the data owner.

Having handled the quasi-identifier attributes, we must also consider the sensitive attributes and how to alleviate the disclosure risks raised in Chap. 2, regarding the weaknesses of variants k-anonymity algorithms such as CG-Anon. We address this issue with a sensitive attribute severity weighting $S(c)$ where $c \in SA$. SA is the list of sensitive attributes and $S(\cdot)$ maps the sensitive attribute category to its weight.

Example 3.2 In Table 3.1, SA denotes the list of diseases (sensitive attribute) and $S(\cdot)$ maps the disease category to its weight, which in this case is simply the guideline illness duration (in time—months, years...) for a given disease. So, $S(\text{Flu}) = 5$ indicates a duration of 5 days. Following this scale, the risk of privacy loss for a tuple containing *"COVID-19"* is higher than for a tuple with *"Mild Cough"*.

3.2.4 CG-Kanon Algorithm

The CG-Kanon scheme uses both the severity weighting and equivalence class formations to hide tuples with highly sensitive values in larger equivalence classes while tuples of lower sensitivity are classified in smaller equivalence classes. For instance, a tuple concerning a "COVID-19" is classified in a 20-anonymity equivalence class while "Flu" could be placed in a lower level equivalence class say, 5-anonymity. This idea of hiding more sensitive values in larger equivalence classes

Table 3.1 Disease severity weightings

Disease	Severity
Flu	3
COVID-19	7
Mild cough	3
Allergic rash	5
Fever	5

does not affect the absolute level of k-anonymity for different sensitive attribute categories. It is instead a relative statement regarding the level of k-anonymity required for different sensitive attributes in the anonymised dataset.

The CG-Kanon algorithm begins by determining a set of quasi-identifiers $Q_T = \{q_T(1), \ldots, q_T(y)\}$. These q_T, are transformed by generalisation and suppression to obtain candidate datasets namely $T'(q_T)$ for each quasi-identifier combination. In order to define the sizes of equivalence classes in which to classify the data for anonymity, we compute an absolute required minimum level of k-anonymity (k_{min}) for the dataset and use k_{min} to guarantee a global minimum level of k-anonymity that all equivalence classes must adhere to in the dataset. k_{min} is computed as follows:

$$k_{min} = \max\left(k_{cons}, \min\left(S_T(\cdot)\right)\right) \tag{3.4}$$

where k_{cons} is a fixed minimum level of k and $S_D(\cdot)$ is the set of all severities for the dataset D. The definition of k_{min} shows that the global minimum level of k-anonymity is fixed at k_{cons} or at the lowest level of attribute sensitivity in the dataset when $\min(S_D(\cdot)) > k_{cons}$. If $k_{cons} = 5$ and $\min(S_D(\cdot)) = 3$ then $k_{min} = 5$. However if $\min(S_D(\cdot)) = 7$ then $k_{min} = 7$ instead. The CG-Kanon scheme uses k_{min} as the k-anonymity baseline when deciding on appropriate equivalence classes for tuples based on severity weighting.

The severity penalty for each classification is computed once a value for k_{min} has been obtained, to enable the CG-Kanon scheme to optimise the information loss and privacy cost-benefit trade-off. The severity penalty determines the privacy loss for a single tuple $t_i \in T(\cdot)$ and is computed as follows.

$$SP_{t_i,j} = \frac{S_{t_i}(c)}{|e_{t_i,j}|} \tag{3.5}$$

where $T(\cdot)$ is the dataset, $S_{t_i}(c)$ is the severity weight of sensitive attribute $c \in t_i$, and E is the set of equivalence classes such that $|e_{t_i,j}|$ is the size of the equivalence class in which a tuple t_i is classified in during the jth iteration of the CG-Kanon scheme.

Example 3.3 From the severity penalty computation, highly sensitive attributes in small equivalence classes result in high penalties and vice versa. So, if a "COVID-19" report with a severity weighting of 20 were located in a 5-anonymity equivalence class, a penalty of $\frac{20}{5} = 4$ is generated. An case of "Flu" with a severity weighting of 5 generates a severity penalty of 1, indicating that this information is comparatively less sensitive. The CG-Kanon scheme uses the severity penalty as a criterion besides, information utility, to determine in which equivalence class to best classify a tuple to minimise the overall sensitive information exposure risk.

Finally, the CG-Kanon scheme computes the aggregate severity penalty, $SP_{tot,j}$, for the entire dataset, to determine whether the obtained anonymised dataset satisfies the predefined threshold goals of privacy and utility. $SP_{tot,j}$ is computed as follows:

$$SP_{tot,j} = \sum_{t_i \in T} SP_{t_i,j} \tag{3.6}$$

and expresses the total severity penalty for the dataset as the summation of the severity penalties of the individual tuples. The $SP_{tot,j}$ is then feed into a fitness function to decide whether each tuple in T satisfies both objectives. We express the fitness function as follows:

$$FF_j^{CG-Kanon} = \frac{1}{\max\left(SP_{tot,j}, LM_{CG,j}\right)} \tag{3.7}$$

So, the result for $FF_j^{CG-Kanon}$ at iteration j is the inverse of the maximum of $SP_{tot,j}$ and $LM_{CG,j}$. Recall that a high $SP_{tot,j}$ indicates a strong risk of privacy exposure, while a high $LM_{CG,j}$ indicates a high level of information loss. Therefore, it is desirable that the fitness function generates results that iteratively converge towards a high value for $FF_j^{CG-Kanon}$, expressed by low values of $SP_{tot,j}$ and $LM_{CG,j}$ respectively. Algorithm 3.1, summarises the description of how the CG-Kanon scheme works.

```
Input    : T, Q_T
Output   : T'
Function Anonymise(T, T', Q_T)
    Q_T ← {q_T(1), ..., q_T(y)};
    for i = 1 to y do
        for ∀q_T(i) ∈ Q_T do
            Apply generalisation and suppression on D ;
        end
        Generate T'(q_T(i)) ; /* - Candidate datasets          */
    end
    k_min ← max(k_cons, min(S'_T(·)));
    for ∀T'(q_T(i) do
        SP_{t_i,j} ← S_d(c) / e_{t_i,j} ;
        SP_{tot,j} = Σ_{t_i∈T} SP_{t_i,j} ;
    end
    FF_j^{CG-Kanon} ← 1 / max(SP_{tot,j}, LM_{CG,j}) ;
EndFunction
for ∀q_T(j) do
    return most fit T'(q_T(j)) ;
end
```

Algorithm 3.1: CG-Kanon

The main drawback of the CG-Kanon algorithm is that, depending on tuple distribution, the diversity of the sensitive attributes in large equivalence classes can be quite low and this negatively impacts on privacy. As well, a large proportion of tuples are suppressed to satisfy the minimum level of k-anonymity which results in high information loss. We addressed this by limiting the size of equivalence classes to a pre-defined threshold size and found that this reduces the number of suppressions to satisfy k_{min}-anonymity. We still have the caveat of inferential attacks and so augment the CG-Kanon scheme with the CG-Diverse scheme (l-diversity algorithm inspired) to help circumvent these attacks.

3.2.5 CG-Diverse Scheme

In CG-Diverse we extend CG-Kanon by using $SP_{tot,j}$ instead, to classify tuples into equivalence classes. The CG-Diverse scheme computes the average severity, AS_T, for T as well as the equivalence class average severity weighting AS_e. The AS_T is computed for T and is used to start the anonymisation process to ensure that the target level of l-diversity in T is such that $l = AS_T$. We compute AS_T as follows:

$$AS_T = \frac{\sum_{t_i \in T} S_{t_i}(c)}{|T|} \tag{3.8}$$

A high AS_T implies a higher level of diversity in the entire dataset. As a stopping criterion for deciding when an acceptable level of k_{min} and AS_T has been satisfied by all the equivalence classes, we bound the l-diversity range with the severity weighting scale and use AS_T to compute the fitness of the dataset with respect to privacy and utility. We employ the following modified fitness function, expressed as follows:

$$FF_j^{CG-Diverse} = \frac{1}{\max\left(AS_{T,j}, LM_{CG,j}\right)} \tag{3.9}$$

However, as mentioned before suppressing the equivalence classes that fail to meet the required levels of AS_T and k_{min} would result in a high level of information loss. Therefore, we alleviate this problem by identifying equivalence classes with a lower average severity (but adequate relative diversity) to avoid high suppression rates. This is achieved by assessing the privacy of individual equivalence classes that do not meet the global AS_T-diversity requirement. To this end the equivalence class average severity weighting AS_e is computed as follows:

$$AS_e = \frac{\sum_{t_i \in T} S_{t_i}(c)}{|e|} \tag{3.10}$$

The AS_e of an equivalence class is compared to the relative diversity l_e, and if $AS_e > l_e$ the tuples in the equivalence class are generalised to the highest possible level to avoid suppression. Alternatively, when the diversity is higher than AS_e no changes are made. We note that this procedure is computationally inexpensive since it simply requires comparing AS_e with the actual observed diversity of the equivalence class. Algorithm 3.2, summarises the operation of the CG-Diverse algorithm.

Input : T, Q_T
Output : T'
Function *Anonymise*(T, T', Q_T)
  $Q_T \leftarrow \{q_T(1), \ldots, q_T(y)\}$;
  for $i = 1$ *to* y **do**
    for $\forall q_T(i) \in Q_T$ **do**
      Apply generalisation and suppression on D;
    end
    Generate $T'(q_T(i))$; /* - Candidate datasets */
  end
  $k_{min} \leftarrow max(k_{cons}, min(S'_T(\cdot)))$
  for $\forall T'(q_T(i))$ **do**
    $AS_{T'(q_T(i))} \leftarrow \frac{\sum_{t_i \in T'(q_T(i))} S_{t_i}(c)}{T'(q_T(i))}$;
  end
  $FF_j^{CG-Kanon} \leftarrow \frac{1}{max(AS_{T'(q_T(i))}, LM_{CG,j})}$;
EndFunction
for $\forall q_T(i)$ **do**
  return most fit $T'(q_T(i))$;
end
for $\forall e \in T'(q_T(i))$ **do**
  $AS_e \leftarrow \frac{\sum_{t_i \in T'(q_T(i))} S_{t_i}(c)}{\|e\|}$;
  ; /* - Ensure Diversity */
end

Algorithm 3.2: CG-Diverse

Example 3.4 Table 3.2 shows the average severity measures calculated for a given sample dataset. The $AS_e = 8$ is calculated as follows: $\frac{5+20+2+3+10}{5}$ using the disease severity weightings given in Table 3.1. By considering Table 3.1, and Eqs. (3.8) and (3.10), the l-diversity range can be restricted to between 3–25, depending on the underlying dataset. Requiring equivalence classes to satisfy the global level of AS_T-diversity might be too restrictive, so we alleviate this issue by moving tuples between equivalence classes to minimise the information loss due to suppression. For instance, in Table 3.2 we observe that in the 5-anonymity equivalence class, "COVID-19" has a severity of 20 which implies a high inference risk. CG-diverse handles such cases by using the AS_e to move the tuple to a more appropriate 20-anonymity equivalence class.

We are now ready to discuss the proposed t-clustering algorithm, which draws on t-closeness principles but uses clustering as the classification mechanism for forming equivalence classes.

Table 3.2 Average severity versus diversity

Age	Disease	Diversity	AS_T	AS_e
30–35	Flu	4	11	8.0
30–35	Fever	4	11	8.0
30–35	Mild cough	4	11	8.0
30–35	Headache	4	11	8.0
30–35	Nose bleed	4	11	8.0
30–55	Ebola	8	11	7.0
30–55	HIV	8	11	7.0
30–55	COVID-19	8	11	7.0
30–55	Lung cancer	8	11	7.0
30–55	Chicken pox	8	11	7.0

3.3 t-Clustering

As mentioned before, the t-closeness approach to privacy enforcement on shared micro-data is performance intensive. We address this problem by using clustering to classify the data to ensure anonymity by t-clustering. t-clustering results in an improved time complexity of $O(m^2 \log n)$ in comparison to $2^{O(n^2)O(m)}$ in the number of operations required for the average case standard t-closeness algorithm.

We incorporate t-clustering into out proposed MOA scheme by ensuring that all the equivalence classes in T are formed by ensuring that the distance between the distribution of sensitive attributes $S(a)$ in each equivalence class and the distribution of $S(a)$ in T is less than or equal to the threshold value t [123]. Building on this, we describe t-clustering as follows:

Definition 3.4 (t-Clustering) Given a micro-data set T, a distance measure $dist(t_i, t_j)$ that describes the similarity between the two tuples t_i and t_j, where t_j is the centroid of a cluster (equivalence class) e_l, and e_l, $S(a_{e_l})$ the entropy of the severity weighting of a sensitive attribute in e_l, and $S(a_T)$ the cross-entropy of the severity weighting of a sensitive attribute in e_l and T. We define t-clustering as a mapping of t_i to e_l, such that $e_l \cap e_y = \emptyset$, and $\|e_l\| \geq k$, such that $\forall t_i, t_j \in e_l$, $Dist(t_i, t_j) \leq Dist(t_i, t_l)$ and $Dist(S(a_{e_l}), S(a_T))) \leq Dist(S(a_{e_y}), S(a_T))$.

Based on this definition of t-clustering, we express the optimal solution to the t-clustering problem as follows:

Definition 3.5 (Optimal t-Clustering) Given a micro-data set T, a distance measure $dist(t_i, t_j)$, and a distance measure $Dist(S(a_e), S(a_T))$ we define optimal t-clustering as a mapping of t_i to e_l, such that in addition to satisfying t-clustering (Definition 3.4, all cluster sizes in T' are maximised, while at the same time minimising $Dist(S(a_e), S(a_T))$, and $dist(t_i, t_j)$.

We ensure t-clustering by computing the minimum cluster size required to guarantee a global minimum level of t-closeness for all clusters. Inspired by the CG-Kanon scheme, the clustering algorithm uses a value k_{min} as the minimum cluster size and moves tuples into appropriate clusters based on the calculated severity weighting as well as the distance measure (distance between a given tuple and the tuple representing the cluster centroid). This ensures that a minimum cluster size will be adhered to by all the clusters created and in addition a minimum level of anonymity for the tuples within each cluster is guaranteed.

Based on the cluster size, we must determine which tuples to either include or exclude from a cluster. As a first step from the definition a clusters, we evaluate the relative distance between tuples to decide on which tuples to classify in the same cluster. In order to compute the inter-tuple distance, we consider both categorical and numerical attributes. The distance between categorical attributes is measured using the Jaccard's coefficient and is defined as follows:

$$sim_{t_i,t_j} = \frac{Q_{t_i} \cap Q_{t_j}}{Q_{t_i} \cup Q_{t_j}} \tag{3.11}$$

where t_i, t_j are two given tuples, Q_{t_i} and Q_{t_j} are the respective quasi-identifiers for t_i and t_j, and t_j represents the centroid of the cluster into which t_i is classified. We note that the value of $\sim (t_i, t_j)$ varies between 0 and 1, with 1 indicating a strong similarity (small) distance between the tuples, and 0 indicating a strong dissimilarity (high distance) between the tuples, based on the quasi-identifier attributes.

However, solely considering attribute similarity as the decision criteria to determine classifications, can result in a high number of outliers (information loss due to misclassification/inability to classify) that get suppressed from the generated anonymised dataset. Suppressions increase the rate of information loss, so we also look at numerical values and compute the Euclidean distance to compare the difference between attributes by using an n-dimensional space function which is represented as follows:

$$Dist(t_i, t_j) = \sqrt{\left(t_i\,(a_1) - t_j\,(a_1)\right)^2 + \ldots + \left(t_l\,(a_m) - t_j\,(a_m)\right)^2} \tag{3.12}$$

where a_m represents a given attribute in Q. Tuples separated by a small Euclidean distance are classified in the same cluster. The next step requires us to consider the sensitive attribute severity weightings. To this end, we compute the average severity weighting AS_T for T and the average severity weighting AS_e for e for a given cluster (equivalence class). We use both values in the same way that t-closeness is used to decide on tuple classifications based on statistical distributions of sensitive attributes, and also to prevent skewness as well as similarity attacks.

We then evaluate the level of loss of privacy with respect to information loss in forming the clusters. This is in line with using the t parameter in the t-closeness scheme as a method of optimising the utility of the dataset, while at the same time enforcing privacy. To this end, we use a fitness function to decide on the value of t to use in the clustering algorithm and express the fitness function as follows:

$$FF_j^{t-clustering} = \frac{1}{Max\left(AS_{T,j}, LM_{CG,j}\right)} \tag{3.13}$$

when $FF_j^{t-clustering}$ is low this implies a high degree of loss of either privacy or information, while a high value is an indication of a reasonably good balance between privacy and data utility.

In order to evaluate the level of diversity of sensitive attributes in e with respect to T, we compute the Kullback-Leibler distance between AS_e and AS_T, to determine whether or not the variation of sensitive attributes in e enforces information utility and privacy. We use the Kullback-Leibler distance (divergence) to provide a measure of entropy which is important in determining the maximum likelihood of inference leading to privacy disclosures due to skewness and similarity attacks. So the Kullback-Leibler distance provides a standard measure for assessing anonymised datasets from the statistical perspective. An added advantage is that because it is not as cost intensive to compute as the Earth mover's distance measure used in the t-closeness scheme[123, 126], we are able to make performance gains in computing the anonymised dataset. We note also at this stage that the added severity weighting metric for the sensitive attributes, compensates for any weaknesses the Kullback-Leibler divergence may represent in terms of accounting for the semantics of the dataset in ensuring privacy. We compute this as follows:

$$Dist\left(AS_e, AS_T\right) = \sum AS_e \log \frac{AS_e}{AS_T} \leq t \tag{3.14}$$

where

$$\sum AS_e \log \frac{AS_e}{AS_T} = H(AS_e) - H(AS_e, AS_T)$$

such that $H(AS_e) = \sum AS_e \log AS_e$ is the entropy of AS_e and $H(AS_e, AS_T)$ is the cross entropy of AS_e as well as AS_T.

When $Dist\left(AS_e, AS_T\right) \leq t$ the anonymised dataset mimics t-closeness by ensuring that sensitive attributes are classified according to severity of exposure. So for example, we would classify $S\left(Cancer\right)$ in a larger cluster and ensure a mixed distribution of sensitive attributes. In this way the chance of identifying users based on similar chronic or serious illnesses is reduced, in comparison to classifying $S\left(Cancer\right)$ in a cluster that is dominated by minor illnesses. This is similar to the reasoning we employed in CG-Diverse. When $Dist\left(AS_e, AS_T\right) \nleq t$, we rerun the algorithm to re-compute cluster structures.

In the next section we provide a complexity analysis of the average case running time for the proposed scheme and compare this to the t-closeness scheme.

3.4 Empirical Evaluation

While we have used examples based on a synthetic healthcare dataset to explain the solution approach, the results we now discuss are drawn from crime data we collected using an experimental platform [191]. We chose to use crime data, because it was generated in a real-life context and presents sensitive attributes such as reported crimes that carry the same sensitivity as the "disease" attribute in the healthcare dataset. Our papers [107, 108], can be consulted for further details.

The experimental platform consisted of a server with an Ubuntu server 12.04 operating system running on a 64 bit machine with 8GB RAM and a processor speed of 3.2GHz (Intel Xeon E3-1230 Quad Core). The dataset consisted of 250,000 records. Table 3.3 provides an example of the composition of the synthetic crime dataset.

Referring to Table 3.3, quasi-identifiers which more closely match the k-anonymity requirement for CG-Kanon were generated before the anonymisation process.

We qualitatively assessed the anonymised data produced by the CG-Kanon and the CG-Diverse algorithms, by considering aspects such as information loss, classification accuracy and the impact of the weighting scheme on linking and inference attacks. The data (see Table 3.3, for example) was transformed by applying an adjustment factor to compensate for the biases that might have been introduced during the data reporting and collection process [39, 107, 108].

We describe bias in our case as the difference between the reported frequency distribution of health records and the real-life frequency distribution. The assumption being that third party health data analytics service providers would be more interested in the true underlying statistical patterns rather than individual values. By employing hypothesis testing a random variable Z_c with a normal ($N(0, 1)$) distribution is described as follows:

$$z_c = \frac{(O_c - E_c)^2}{E_c}$$

where E_c is the expected frequency, and O_c is the observed frequency.

Table 3.3 Sample dataset

Firstname	Lastname	Age	Gender	Postcode	Disease
Original data					
Heather	Cornish	30	Male	34520	Theft
Andrew	Bale	34	Male	34530	Murder
Pretty	Simpson	31	Male	34540	Burglary
Virginia	Dyler	30	Male	34544	Illegal gambling
Andrew	Ellison	34	Male	34545	Fraud
Alan	Butler	34	Male	34545	Drunken driving

Across all crime categories this translates to a χ^2 random variable which is used for goodness-of-fit. The adjustment factor for each crime category can then be evaluated using: $adj_c = \sqrt{z_c}$. Table 3.4) provides an example of adjustment factors calculated for corresponding crime categories.

One aspect that is central to our proposed MOA scheme is that of quasi-identifier weightings which are used during the anonymisation process to determine the cost benefit tradeoff with respect to privacy and information loss. For instance, for a quasi-identifier containing the attributes *Age*, *Suburb*, and *Reporter*, we would denote this as $A_{w_{Age}} : S_{w_{Suburb}} : R_{w_{Reporter}}$ (Fig. 3.5).

When equal weights were assigned to the quasi-identifier attributes we denoted this as: denoted as an $A1 : S1 : R1$. In a similar fashion, we use $A10 : S5 : R1$ which would indicate weights of 10, 5, and 1 assigned to each of the attributes. So applying this to the attributes *Age*, *Suburb*, and *Reporter* in a given quasi-identifier would indicate that the attributes were each weighted at 10, 5, and 1 respectively. We also set $k_{constant}$ was set to 5 for all results on CG-Kanon anonymisation mainly as a minimum required anonymisation level.

In terms of increasing the level of sensitive attribute diversity, if the minimum crime severity level for the data was set to 3 and $k_{min} = 5$. For CG-Diverse, we set the lowest diversity level to 3 and applied this to all anonymisations as the

Table 3.4 Sample data distributions and adjustment factor: An example

Crime	Generated data	Real distribution	Z_c	adj_c
Corruption	555	650	13.88	4
Rape	574	600	1.13	1
Disorderly conduct	561	500	7.44	3
Drug use	729	760	1.26	1
Murder	777	300	758.43	28
Theft	699	1700	589.41	24

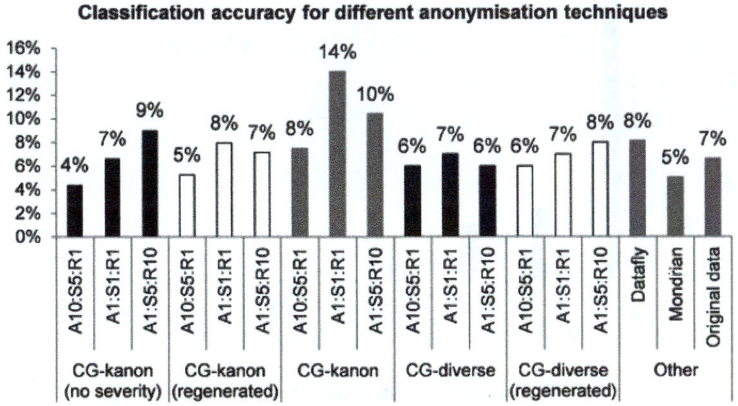

Fig. 3.5 Classification accuracy of CG-Kanon and CG-Diverse [108]

base minimum privacy level. Typically the lower severity crimes (less sensitive attribute values) were located in such equivalence classes, so this was acceptable. For consistency and statistical validity, all algorithms were allowed to run for 30 minutes before their execution was terminated. The reason for terminating the execution also goes in line with our goal of maximising the processing and power resources on the device or machine.

Pre-experiment sampling indicated that running for shorter periods, say 15 minutes resulted in high severity penalties and high information loss for larger equivalence classes. By contrast, running for 30 minutes resulted in only between 3–6% of tuples meeting the minimum anonymity level. As we were aiming for higher numbers of tuples meeting the minimum anonymity level, we ran the experiment until a point of saturation, where the further classification success rates came at the cost of processing time efficiency. We recall that our primary goal is to ensure our proposed solution works efficiently on low powered and low processing devices.

On termination of the anonymisation procedure, we checked the anonymised data for compliance with the desired level of privacy. The tuples that failed to satisfy the privacy criteria were resubmitted to the processing procedure for further transformation. Figure 3.6 shows how the CG-Kanon algorithm classifies data with respect to equivalence classes when no severity weightings are applied. In this case, crimes are clustered around smaller sized equivalence classes which results in high levels of information loss due to the transformations needed to ensure privacy. The transformation procedure is also processing intensive.

The severity penalty is applied, to address the issue of information loss that arises with the CG-Kanon scheme. As shown in Fig. 3.7 more severe crimes (high sensitive information) are classified in larger equivalence classes. This reduces the level

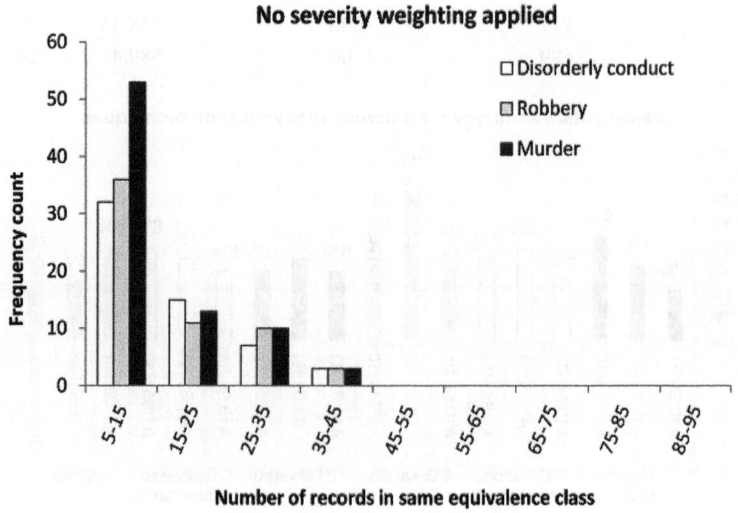

Fig. 3.6 Severity impact on dataset (no severity weighting) [108]

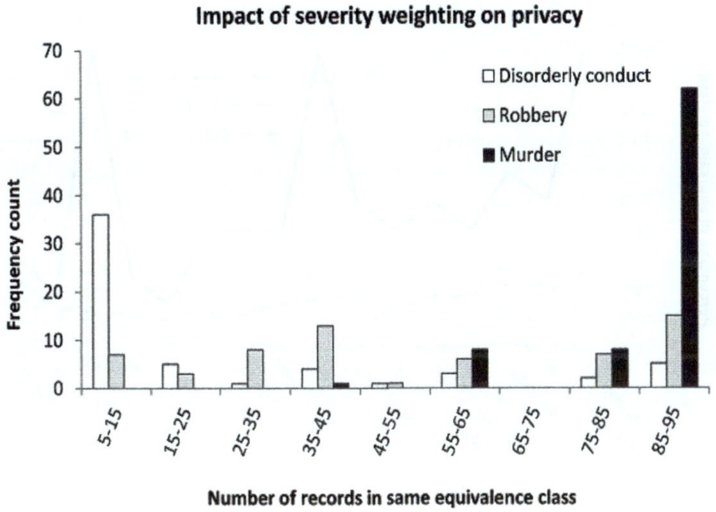

Fig. 3.7 Impact of severity weighting on privacy [108]

of information loss, but introduces the potential for inferential disclosure. Similar to our discussion in Chap. 2, regarding adversarial background knowledge on the mechanisms applied to anonymise the data. For instance, from Fig. 3.1 an adversary might be able to guess that more severe crimes have a higher frequency in certain regions or suburbs with say, *"Murder"* being as high as 31%. We discussed the impact of monotonous sensitive attribute value distributions on disclosure potential in Chap. 2, and so we address this with the CG-Diverse scheme.

As shown in Figs. 3.8 and 3.9, based on the $A1 : S1 : R1$ weighting and an average severity level of 11, we study the frequency of sensitive attributes for both CG-Kanon and CG-Diverse. In this case, the average diversity in CG-Kanon varies between 10 and 30% while that of CG-Diverse is much lower at 9% to 14%. By using the severity weightings we are able to distribute highly sensitive records into equivalence classes that make distinguishing individual record lower. The desired lower frequency (i.e. higher diversity) for more severe crimes is evident in CG-diverse whereas in CG-Kanon this is not the case. Diversity in CG-Kanon relies primarily on using severity disclosure risk to more tuples to equivalence classes that reduce the disclosure risk. More severe crimes (Rape and Murder) in this case actually have lower average diversity and consequently less risk of inferential exposure.

The deviation from the mean frequency also becomes lower for more severe crimes as severity increases. In the sense that the average diversity increases with crime severity but the variance decreases indicating that more severe crimes will be less vulnerable to inference attacks.

Since l-diversity guarantees at least k-anonymity when $k = l$. The average diversity is also considered over both schemes (CG-Kanon, and CG-Diverse).

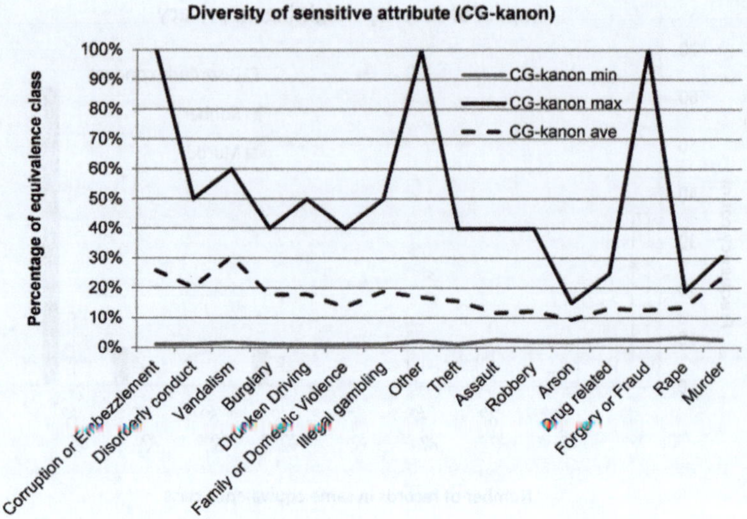

Fig. 3.8 Sensitive attributes frequency for CG-Kanon using A1:S1:R1 [108]

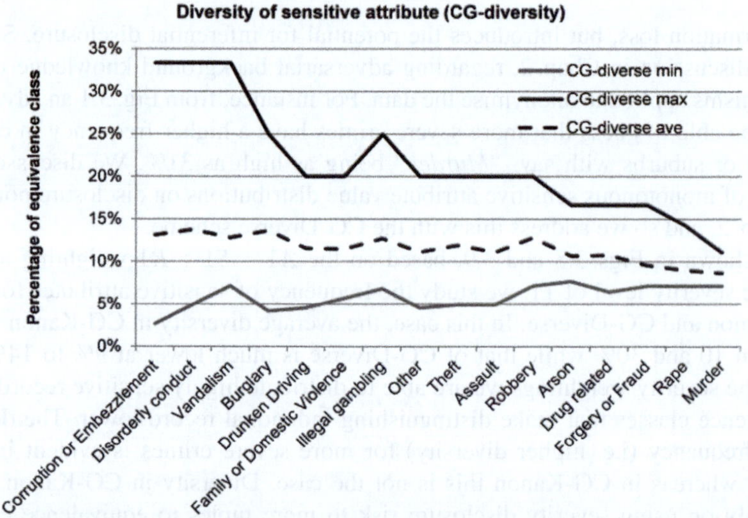

Fig. 3.9 Sensitive attributes frequency for CG-Diverse using A1:S1:R1 [108]

In Fig. 3.10, shows a comparison of sensitive attribute frequency across the three proposed schemes namely: CG-Kanon, CG-Diverse, and *t*-clustering.

The lowest diversity of 3 may appear weak from the privacy perspective when compared to the global diversity of 11 but it is unlikely, practically speaking, that severe crime (sensitive data) will be included in such lower diversity equivalence classes. For instance, if the earlier results obtained for the CG-Kanon algorithm are

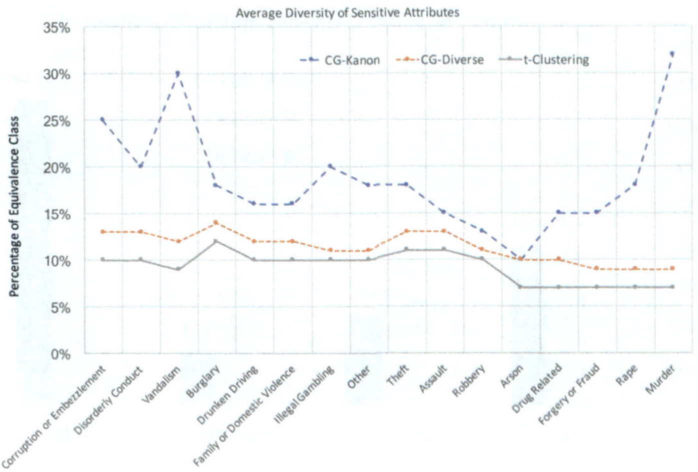

Fig. 3.10 Sensitive attributes frequency: CG-Kanon, CG-Diverse, and t-clustering [108]

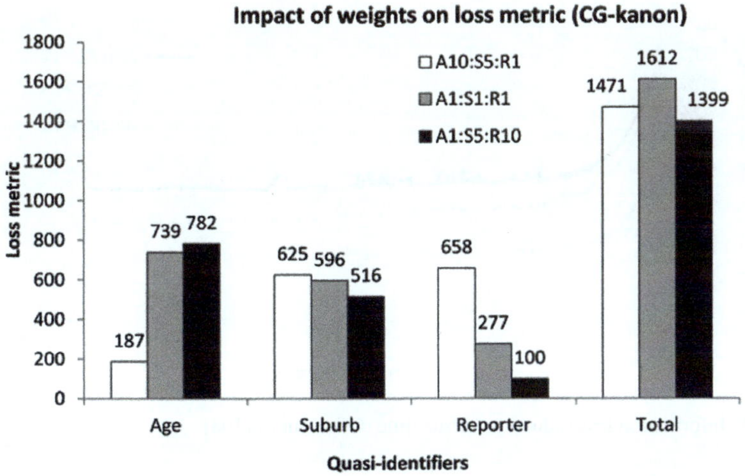

Fig. 3.11 Information loss for CG-Kanon [108]

re-visited for the most serious crime ("Murder") was in an equivalence class of size 90 and still only achieved a 3-diversity.

Figures 3.11 and 3.12 show the aggregated information losses for different weighting schemes after termination of the algorithm. We selected three weighting schemes to monitor how the algorithms perform when attributes with varying granularity are weighted differently. For instance the $A10 : S5 : R1$ scheme overweights the *Age* attribute which is highly granular and under weighs the *Reporter* attribute, while $A1 : S5 : R10$ test the opposite scenario and $A1 : S1 : R1$ is equivalent to having no weighting scheme. The marginal increase in

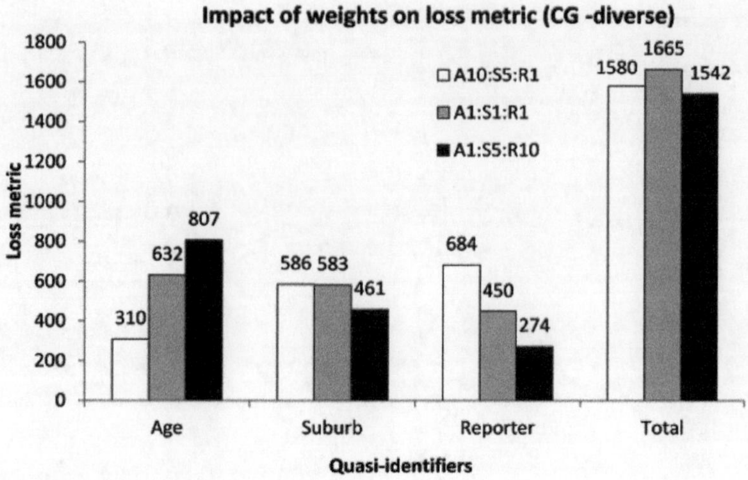

Fig. 3.12 Information loss for CG-Diverse [108]

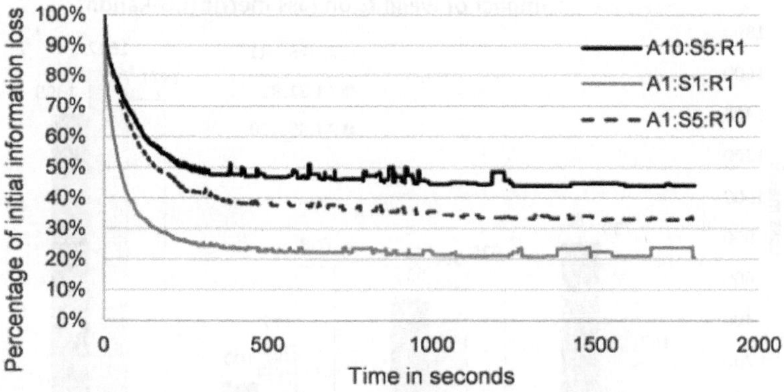

Fig. 3.13 Information loss reduction versus time (CG-Kanon) [108]

information loss for CG-diverse relative to CG-Kanon seems quite acceptable given the improved privacy provided by CG-Diverse. For the results the information loss across the three weighting schemes was on average 7% higher for CG-diverse. However, this reduced data utility is acceptable given the need for better data privacy.

One further insight relates to the number of parameters that are used for the fitness function in selecting quasi-identifiers. Figures 3.13 and 3.14 indicate that information loss for CG-diverse proportionately lower than that of CG-Kanon. We attribute this to the fact that CG-Kanon searches for solutions that minimise both the information loss and the severity penalty, in addition to satisfying k-anonymity whereas CG-diverse only focuses on minimising information loss and aim to meet the diversity requirement. The severity penalty in CG-Kanon increases the search

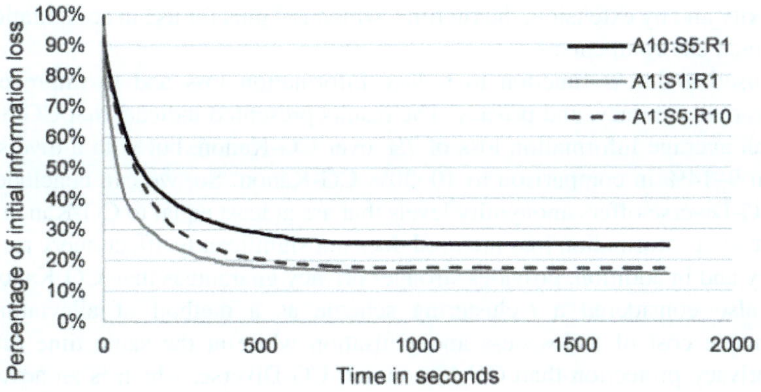

Fig. 3.14 Information loss reduction versus time (CG-Diverse) [108]

space and reduces the efficiency of the algorithm. For instance, at termination the reduction in the initial information loss for $A10 : S5 : R1$ in CG-diverse was 74% compared to 55% for CG-Kanon. For an average cluster size (equivalence class) of 7, information loss was on average 14.6% mainly because the t-clustering algorithm relies on the k-anonymisation principle to bucketise tuples and only seeks to establish t-closeness if the observed diversity in an equivalence class is less than t the threshold for diversity with respect to the entire dataset.

Furthermore, based on the principles of l-diversity, the costs of clustering grow linearly with the size of the dataset.

3.5 Discussion

In this Chapter, we discussed three methods of anonymising data on low powered and low processing devices. To this end, we presented modified k-anonymity, l-diversity, and t-closeness anonymisation schemes namely, CG-Kanon, CG-diverse, and t-clustering. The schemes presented operate by augmenting the standard k-anonymity, l-diverse, and t-closeness algorithms by enabling automatic classification and anonymisation of data.

For consistency, we discussed the schemes with examples based on our running case study of health data. However, for our experiments we used crime data that was drawn from a real-life experimental project. The reason for doing so was to access the effectiveness of our proposed schemes in a real-life scenario rather than on a synthetic healthcare dataset. Furthermore, our crime data contains a large volume of real sensitive data and is vulnerable to linking and inferential attacks. To match privacy with utility, we used a random sampling approach without replacement so, historical released reports were excluded from being selected in subsequent releases. The sampling approach also offers the advantage of reduced computational complexity and by extension, the runtime, which is a plus for use in computationally constrained environments.

We used a fitness function to reduce information loss and to improve both classification accuracy, and privacy. The results presented indicate that CG-diverse incurs an average information loss of 7% over CG-Kanon, but with a diversity of between 9–14% in comparison to 10–30% CG-Kanon. So, we can conclude that, since CG-Diverse offers anonymity levels that are at least equal to CG-Kanon's, the percentage of information loss incurred does not significantly affect query response accuracy and in addition, provides stronger privacy guarantees than CG-Kanon.

We also considered a t-clustering scheme as a method of alleviating the performance cost of t-closeness anonymisation while at the same time offering better privacy protection than CG-Kanon and CG-Diverse, which is an advantage for low resource settings. Basically, what we did was build on the method of ranking sensitive attributes by a severity weighting proposed for the CG-Kanon and CG-Diverse scheme. The ranking system was used to classify tuples to enforce diversity, so that we do not always have to compute statistical distributions to satisfy t-closeness. The tuples are classified to minimise the risk of privacy disclosure of tuples containing high severity weight sensitive attributes. Clustering has the advantage of reducing the need for extensive attribute generalisation in order to classify tuples based on similarity. This is good, in addition, because it reduces the cost of information loss. As we have mentioned earlier, high levels of information loss make datasets unusable in practical situations. By considering severity weightings both for individual clusters and the entire dataset, we mimic the t-closeness principle, of seeking to distribute tuples in ways that ensure that the difference in distributions both within the equivalence classes and the entire dataset, does not surpass a threshold value of t. In this way, the proposed schemes also offer protection against skewness and similarity attacks. Finally, a further benefit of the schemes we have presented in this chapter, is that because they are is not performance intensive, they can be used on low-powered, low-processing networks for guaranteeing privacy of data in various environments.

In the next Chapter, we place a heavier emphasis on quasi-identifier discovery as an added measure to guaranteeing privacy. We go on the premise that discovering all quasi-identifiers in a dataset is integral to guaranteeing privacy and present an algorithm to enable efficient quasi-identifier discovery in large high dimensional datasets. Large high dimensional datasets arise from data composition scenarios such as the one we presented in Fig. 3.1.

Chapter 4
High-Dimensional Data: Privacy Considerations

In Chap. 3, we discussed a multi-objective anonymisation approach to generating privacy preserving data in low computational and power resource settings and explained that such a mechanism is useful in creating generic privacy-preserving datasets for contexts in which per-query anonymisation is impractical either due to performance or real-time constraints. multi-objective anonymisation is however not well-suited to anonymising high dimensional datasets that result from composing data belonging to different data owners and that can also arise from distributed data sources [177]. One of the reasons for this is that in large high-dimensional datasets, attribute-values are relatively sparsely distributed with respect to the number of attributes [10, 14], and as such when approaches such as multi-objective anonymisation are applied to large high dimensional data, the resulting anonymised dataset incurs high levels of information loss. As mentioned before, high information loss in anonymised datasets is undesirable for data analytics operations.

In this Chapter, we present an approach to transforming large high dimensional datasets to ensure high privacy with low information loss [171, 172, 172]. Our proposed solution rests on the fact that efficient quasi-identifier discovery is crucial to enabling efficient anonymisation. The aim is for our quasi-identifier discovery mechanism, to support third-party data anonymisation providers in sharing datasets in a privacy preserving way (see Fig. 3.10).

In order to identify quasi-identifiers, a three-step process is employed: (1) A cardinality metric is used to determine explicit identifiers (attributes that directly identify individual records) based on the frequency of occurrence of unique attribute-values. Explicit identifiers are suppressed from the dataset to remove the risk of disclosure; (2) The remaining attributes are evaluated using a quasi-identifier discovery algorithm to determine attribute-value combinations that are likely to enable de-anonymisations. (3) The identified quasi-identifiers are evaluated to determine which attribute combinations can be placed together safely without the risk of disclosure.

A. V. D. M. Kayem, *Enabling Privacy Preserving Data Analytics*, Advances in Information Security 92, https://doi.org/10.1007/978-3-031-93906-8_4

An empirical study on large high-dimensional data drawn from sample healthcare (Synthetic data), fitness (CrossFit), and Twitter (social media) indicate that our solution is scalable in terms of performance and offers an effective means of protecting against de-anonymisations of large high-dimensional datasets. Results of further optimisations using GPU computation mechanisms are accessible in [176].

4.1 High-Dimensional Data

Large volumes of high dimensional data are generated in a wide variety of applications ranging from healthcare, online marketing, to education, for instance. Service delivery in these applications can benefit greatly from the information drawn from analysing the data. For instance, online marketing platforms use purchase data to determine buying trends that support services such as advertising, product recommendations, and even stock management. Such large high-dimensional data sets are particularly useful from the data analytics perspective because the dimensions of the data offer a rich source of information from which useful inferences can be drawn. For instance, a healthcare data set that contains genomic information is more useful for drug discovery research than one that contains solely information regarding patient vital signs. In order, to comply with data privacy and ethical regulations, an important and necessary pre-processing step is to ensure that the data is privacy-preserving.

However, as mentioned before, anonymising large high-dimensional data sets, using syntactic data anonymisation mechanisms such as multi-objective anonymisation, is impractical because the anonymisation process provokes a high level of information loss. The reason for this is threefold: (1) Many dimensions may be noisy and irrelevant for anonymisations which raises the likelihood of classifying certain attributes and quasi-identifiers as similar depending on the dimension of the data that is considered, and dissimilar when alternative dimensions are considered [24, 26, 56, 233]; (2) When represented using a relational data model, high dimensional data is characterised by a large number of row and columns, which as the data grows in scale can result in state space explosion issues for the anonymisation algorithms [29]; (3) Distinguishing between sensitive and non-sensitive attributes is challenging [215]. This in the sense that our discussion in Chap. 3 highlights the fact that sensitive features (attributes) alone are the main reason for successful de-anonymisations. Furthermore, creating an exhaustive collection of sensitive and non-sensitive features (attributes) is difficult to do "correctly" for high-dimensional datasets in which distinct patterns grow with the amount of data collected about an individual [70, 115, 203]. For instance, in genome datasets attributes that contain information on human genetic phenotypes are challenging to transform without sacrificing privacy in favour of information availability, and vice versa.

The problem with which we are faced therefore, is that of generating low-information loss anonymised high-dimensional data for scenarios in which sub-optimal privacy is acceptable, and in which the use-case scenarios lend themselves well to the application of syntactic data anonymisation mechanisms.

4.2 Ensuring Low Information Loss

In order to address the issue of high information loss due to syntactic anonymisation of high-dimensional data, we present an attribute compartmentation approach to transforming the data. Attribute compartmentation works as a three-step process. First, we classify attributes as either explicit or quasi-identifiers. Explicit identifiers are ones that can be used for one-to-one de-anonymisations. For instance, a *Name*, *Social Security Number*, or *IP-Address* could be used to identify an individual uniquely. As mentioned before, quasi-identifiers by contrast, are attribute combinations with values that can be used for de-anonymisations. Individually, such attributes pose no threat, only when used in combination, does the threat appear. For instance, *Age, Gender, Postcode* could be used as shown in Chap. 2 (see Sect. 2.4) to de-anonymise a user, but individually neither attribute *Age, Gender* nor *Postcode* could be used to do so successfully.

We go with the premise that the explicit personal identifiers are essentially personal identifying information and therefore exhibit a high degree of unicity in datasets. This property lends itself well to enabling direct inferences of hidden sensitive data associated with the affected tuples. Once identified, the affected attributes (explicit identifiers) are suppressed from the dataset to remove the risk of one-to-one mappings to individual records in the dataset.

The remaining attributes in the dataset are then evaluated to determine which attribute-value combinations (quasi-identifiers) enable inferences that can be exploited to discover hidden sensitive information. Similar to the case of explicit identifiers, we evaluate attribute combinations for unicity. However, in this case, instead of suppressing the attribute combinations, we seek instead to do either one of or both of the following: (1) We apply transformation measures to the data within the identified attributes to reduce or remove the degree of unicity. We discussed data transformation measures in Chap. 2 and will delve into this in more depth later on in this Chapter. (2) We create silos (or compartments) of data to break linkability or inferential risk. This operates along the principle of identifying composite keys within the data, and creating subsets of the data to ensure that no given subset contains an attribute combination that forms a composite key. This second approach lends itself well to minimising information loss because the creation of such subsets of data implies that the data can be published in a privacy-preserving way without suppressing the attribute combinations, or applying transformation techniques such as suppression. So we can determine which attributes can be placed together safely in a given dataset.

In the following Section we provide details on how our method works to achieve low information loss on high-dimensional datasets, while guaranteeing privacy.

4.3 Selecting Explicit Identifiers

As a first step to transforming data to ensure anonymity, we identify and eliminate all explicit identifiers.

Remark 4.1 We assume that the data is pre-processed to eliminate all explicit personal identifiers, since these attributes are ones that can be used to directly identify individuals.

For the rest of this Chapter, we assume a relational database model in which each record appears only once and is described by a series of attributes. The database is represented using an $\mathcal{M} \times \mathcal{N}$ matrix where $\mathcal{M}$ represents the number of records and $\mathcal{N}$ the number of tuples in the database.

We define explicit personal identifiers as follows:

Definition 4.1 (Explicit Personal Identifiers) Let A be a set of attributes such that $A = \{a_1, \ldots, a_n\}$, where each attribute is a function $a_i : R \longrightarrow V$ that maps a set R of realisations of a_i to a set V ($V = \{v_1, \ldots, v_m\}$) of values. An attribute a_i is termed an explicit personal identifier (EPI), iff a_i is injective. That is, $\forall v_j, v_k \in V$: $a_i(v_j) = a_i(v_k) \Longrightarrow v_j = v_k$.

In order to discover the attributes that satisfy the explicit personal identifier definition, we compute the cardinality c of unique values on a per attribute basis and compare this to a pre-defined threshold value t. For instance, if we set $t = 0, 33$ and $c > 0, 33$ this would imply that at least every third attribute value is unique, and that when this threshold is met the attribute under consideration must be classified as an explicit personal identifier. More formally, we define the cardinality of an attribute as follows:

Definition 4.2 (Cardinality) The cardinality $c \in \mathcal{Q}$ of an attribute is given as follows:

$$c = \frac{\text{number of unique values}}{\text{total number of attribute values}} \tag{4.1}$$

Once identified, the explicit personal identifiers are suppressed from the dataset so that no direct and bijective linkages can be used to link the data remaining in the dataset and the suppressed attributes.

In order to determine how we can quantify the quality of the threshold (i.e., the number of unique values required per attribute to classify it as a 1st class identifier), we compute the level of information gain. Information gain indicates how likely it is that the attribute under consideration will or could be exploited for information disclosure.

We do this by first computing the entropy E of the attribute, to determine how much variance the attribute-values have. For instance, an attribute that has the same value for each record (tuple) would have a low entropy and the reverse for the case in which every single value is unique. We calculate information entropy, E, for a dataset with X attributes that have been classified as explicit personal identifiers as per the cardinality evaluation (see Definition 4.2), as follows:

$$E = -\sum_{A}^{X} p_A \cdot \log_2 p_A \tag{4.2}$$

where p_A is the probability of randomly selecting an attribute value, belonging to attribute A, that successfully enables inference of hidden sensitive information associated with the tuple containing the attribute value.

Second, to quantify the quality of the cardinality threshold, and also to deal with inferences due to value similarity (homogeneity) [139]. Quantifying the cardinality threshold is important because we need to know whether or not the threshold allows us to effectively identify explicit personal identifiers without causing information loss by misidentifying explicit personal identifiers. Misidentifying explicit personal identifiers can result in high information loss thereby reducing the value of the dataset for analytics or querying. However, incorrectly identifying explicit personal identifiers can result in information exposure (privacy violations). So we need to maximise the chances of correctly selecting explicit personal identifiers and minimise the cost of information loss.

The measure we use to achieve this is information gain, which is simply the expected reduction in entropy of the dataset D, caused by removing the attribute that has been classified as an explicit personal identifier. The information gain $Gain(D, A)$ of an attribute A relative to a data set D is defined as follows:

Definition 4.3 (Information Gain) Let D be a data set and A_i an of attribute classified as an explicit personal identifier. The information gain of A_i relative to D is computed as follows:

$$Gain(D, A_i) = Entropy(D) - \sum_{v \in V} \frac{|D_v|}{|D|} \cdot Entropy(D_v) \tag{4.3}$$

where V are all the possible values for attribute A_i and D_v is the subset of D for which attribute A_i has value v

The goal of computing information gain is to evaluate which explicit personal identifiers to eliminate from the dataset and which to retain, based on the information gain achieved by retaining these attributes.

Finally, we must consider attributes that have low entropy. Attributes that have a low entropy can be vulnerable to homogeneity attacks [123], which are a variant of semantic attacks (see Chap. 2). Homogeneity attacks were raised by Machanavajjhala et al. [139] as an example of de-anonymisation vulnerabilities that k-anonymised datasets face. Homogenity attacks exploit the semantic similarity of attribute values to infer information about a dataset. For instance, the *Disease* attribute in Table 4.1 that lists ‘COVID-19’ as the value for all records associated

Table 4.1 Original dataset: Adults and corresponding diseases

Name	ID-Number	Age	Gender	Postcode	Disease
Alice	1827–88	34	Female	34521	COVID-19
Bob	1727–88	32	Male	34530	COVID-19
John	1526–98	33	Male	34540	COVID-19
Jeremy	5555–88	37	Female	34551	Lung cancer
Jamie	6654–78	38	Female	34570	Gallstones
James	2345–66	36	Female	34580	Malaria
John	7896–54	42	Male	34555	Thyphoid
Jamie	6789-23	43	Male	34560	Lung cancer
James	8967-23	44	Male	34525	Stomach cancer

with (Alice, Bob, and John), has a low entropy. However this still allows an adversary to infer that all the members in the dataset suffer from an acute respiratory infection, namely 'COVID-19' [139].

We measure inference by computing the relative entropy (Kullback-Leibler Divergence) to determine the level of information gain obtained when comparing the attribute values for inference. Determining the level of relative entropy, provides a measure of the likelihood of the attribute contributing to inference based on the values it contains. We define relative entropy more formally as follows:

Definition 4.4 (Relative Entropy Kullback-Leibler Divergence) Let p and q denote two discrete probability distributions of attribute values. The Kullback-Leibler divergence or relative entropy E_{rel} of p with respect to q is: $E_{rel} = \sum_i p(i) \cdot \log(\frac{p(i)}{q(i)})$ where $p(i)$ is representative of the distribution of values in the dataset without the identified explicit personal identifier(s) values, i.e. $D - (a_i)$ and $q(i)$ represents the distribution of values in the original dataset D.

However, using combinations of attributes (quasi-identifiers) from the remaining attributes in the dataset can still be used for disclosure [212]. In the following section, we propose a method of identifying these attribute combinations (quasi-identifiers).

4.4 Discovering Quasi Identifiers

We augment the standard definition of quasi-identifiers to include sensitive attributes. So that instead of distinguishing between sensitive attributes and quasi-identifiers, we consider that sensitive attributes are in fact a special case of quasi-identifiers with the power to provide more information than that ascribed to standard quasi-identifier attributes in previous works such as [213, 215] and also as discussed in Chap. 2. A further reason is that, defining an exhaustive set of sensitive and non-sensitive attributes is impractical for high dimensional datasets where unique patterns emerge and increase with the volumes of data collected on individual records (tuples). More formally we re-define our augmented quasi-identifier as follows:

Definition 4.5 (Augmented Quasi-identifiers) Let $\mathcal{A}$ be a set of attributes such that $A = \{A_1, \ldots, A_n\}$ and $B := P(A) = \{Q_1, \ldots Q_k\}$, its power set (i.e. the set of all possible attribute combinations). A subset of attributes $Q_i \in Q$ is an augmented quasi-identifier, if and only if B_i, can be used to identify at least one record uniquely and all the attributes $A_i \in Q_i$ are not explicit personal identifiers.

We determine augmented quasi-identifier candidates by evaluating the all of the possible attribute combinations of size $1 \leq l \leq n$, that are likely to be exploited for disclosures of individual tuples (i.e. resulting in re-identification). In order to evaluate the number of attribute combinations that must be considered for any given dataset, a binomial coefficient can be used as follows:

$$C(n,l) = \binom{n}{l} = \frac{n!}{l!(n-l)!} \tag{4.4}$$

where n is the total number of attributes left in the dataset after the explicit personal identifier attributes have been suppressed; l is the number of attributes in a given augmented quasi-identifier under consideration.

We can optimise, by evaluating only attribute combinations without repetitions and without permutations (i.e. the order of the attributes does not matter). In this case, the total number of attribute combinations to consider for augmented quasi-identifier eligibility becomes:

$$\sum_{l=1}^{n} \binom{n}{l} = \sum_{l=1}^{n} \frac{n!}{l!(n-l)!} = 2^n - 1 \tag{4.5}$$

So essentially, $O(2^n)$ potential attribute combinations must be evaluated to determine whether or not they satisfy they criteria for augmented quasi-identifiers. This is a computationally intensive process particularly for large high-dimensional datasets, since the total number n of attributes is high ($n \geq 100$).

We note here that the procedure for deciding on augmented quasi-identifiers is similar to that employed for determining explicit personal identifiers. Similarly, a threshold is decided for the number of unique attribute-value combinations needed to qualify an attribute-combination as an augmented quasi-identifier. However, instead of opting for total suppression of the augmented quasi-identifier attributes as we did for the explicit personal identifiers, we employ data transformation measures (e.g. generalisation, suppression, and perturbation) to increase the degree of similarity (i.e. to reduce unicity).

Since evaluating attribute combinations for augmented quasi-identifier eligibility is a processing intensive operation, discovering augmented quasi-identifiers efficiently is useful in reducing this processing time. We recall, that the potential number of attribute combinations to evaluate is in $O(2^n)$ and that from our discussion in Chap. 3, a primary objective of our proposed anonymisation framework (see Fig. 3.1), that to enable low powered/ low processing anonymisation.

4.5 Optimising the Search for Quasi Identifiers

Assessing attribute combinations for augmented quasi-identifier eligibility requires that all composite keys (unique column combinations) be discovered [2, 169]. As mentioned before, this is a time and processing intensive procedure, so we seek instead to discover the set of minimum unique attribute combinations (minimal augmented quasi-identifiers) needed to discover all the augmented quasi-identifiers in a given dataset. The idea is that the identified minimal augmented quasi identifiers, will be represent attribute combinations that are subsets of the parent attribute combinations. That is by identifying a minimal augmented quasi-identifier, we can determine what the parent quasi-identifier is.

We define a *Minimal augmented quasi-identifier* as follows.

Definition 4.6 (Minimal Augmented Quasi-identifier) An augmented quasi-identfier $B_i \in \mathcal{P}(A)$ is minimal, if and only if B_i contains no combination of attributes B_j such that $B_j \subset B_i$ and B_j is an augmented quasi-identifier.

Theorem 4.1 *The set of minimal augmented quasi-identifiers is sufficient to prevent re-identification.*

Proof Since all the supersets of the augmented quasi-identifiers B in the resulting set of augmented quasi-identifiers are correct, the set of augmented quasi-identifiers is complete. That is, there is at least one augmented quasi-identifier, B_i, such that $B_i \subseteq B$. Furthermore, if a minimal augmented quasi-identifier is valid, then the associated quasi-identifier must also be minimal with respect to the resulting set of minimal augmented quasi-identifiers. Therefore, terminating the search algorithm early does not result in incomplete results. It suffices to discover at least one minimal augmented quasi-identifier to discover all associated attributes that can be exploited for disclosure.

4.5.1 Pascal Triangle Heuristic

In order to compute the set of minimal augmented quasi-identifiers efficiently, we a heuristic based on the notion of the Pascal's triangle. Essentially what we do is to build a Pascal triangle based on the number of attributes under consideration for augmented quasi-identifiers. We employ the following equation to build the Pascal's triangle:

Fig. 4.1 Corresponding Pascal triangle for 4 attributes

Level 0: 1

Level 1: 1 1

Level 2: 1 2 1

Level 3: 1 3 3 1

Level 4: 1 4 6 4 1

Table 4.2 Extract showing positions of Pascal triangle values at Level 4

Pos. 0	Pos. 1	Pos. 2	Pos. 3	Pos. 4
1	4	6	4	4

$$\binom{n}{l} = \frac{n!}{l!(n-l)!} \tag{4.6}$$

where n is the number of attributes under consideration for augmented quasi-identifier eligibility and l is the number of attribute combinations ($>$ 1), without repetition, that can arise from the attributes. For example, for $n = 4$, we have four attributes, say A, B, C, D which will result in $l = 6$ combinations: AB, AC, AD, BC, BD, CD.

To determine which attribute combinations to consider without dealing with repetitions (e.g. AB and BA instead of just AB) , one approach is to use a Pascal triangle. With the Pascal triangle heuristic, instead of evaluating 2^n combinations we aim to reduce the number of computations by half to 2^{n-1}, even in the worse case. For example, considering the set of four attributes A, B, C, D, we aim to determine which ones qualify for evaluation as augmented quasi-identifiers.

Example 4.1 Figure 4.1, shows the resulting Pascal triangle for the four (4) attributes, obtained based on Equation 4.6. In order to determine the augmented quasi-identifier status for the attribute combinations, we have to consider attribute combinations of sizes 2..4. We begin with combinations of size 2. Using the Pascal triangle above, since we know that we have 4 attributes, we look at the values at level 4 and furthermore since we require combinations of size 2, we can determine by observation of the value at Position 2 (see Table 4.2), that we must evaluate 6 combinations. The procedure is repeated for combinations of size 3 and 4, resulting in 4 and 1 combinations respectively to consider for both cases. So, instead of evaluating $((2^4 = 16) + (3^4 = 81) + (4^4 = 256))$ combinations, by using the Pascal triangle heuristic we can reduce the number of evaluations to $(6 + 4 + 1)$ combinations which offers a significant gain in runtime performance.

4.5.2 Building a Powerset Lattice

Once we have the number of attribute size combinations to evaluate, we must proceed to determine what these combinations are. We do this by determining for each attribute combination size, all unique combinations (with no repetitions) or redundancies (i.e. same attributes but different ordering). For example, for four (4) attributes say,

(Age, Gender, Profession, Postcode)

that we denote as follows: (Age) as A, (Gender) as B, (Profession) as C, and (Postcode) as D, we have the following:

- **Attributes of Size 2:**
  - $\{AA, AB, AC, AD, BB, BA, BC, BD, CC, CA, CB, CD, DD, DA, DB, DC\}$
  - By eliminating redundancies and repetitions we have: AB, AC, AD, BC, BD, CD which is a set of size 6 (as deduced by the Pascal triangle heurlstic).
- Repeating the procedure for **Attributes of size 3 and 4** we get the following sets: ABC, ADC, ABD, BCD and $ABCD$ of size 4 and 1 respectively.

These combinations can then be used to build a powerset lattice which is useful in identifying the set of minimal augmented quasi-identifiers. We discuss this in the next section.

4.6 Minimal Augmented Quasi-Identifiers

We must now select the set of minimal augmented quasi-identifiers (quasi-identifiers). This is done by running a minimality test which essentially determines which of the attribute combinations have the highest entropy (high likelihood of exposure). As mentioned before, this is evaluated in the same way as was done for explicit personal identifiers (see Sect. 4.3). The procedure is applied iteratively to the attribute-combinations of size $2 \ldots n$ in the powerset lattice. The set of minimal augmented quasi-identifiers is complete if each of the attribute combinations in the set is a subset of at least one of the augmented quasi-identifiers in the powerset lattice. Example 4.2 helps to illustrate the process of forming the set of minimal augmented quasi-identifiers.

Example 4.2 Suppose the powerset lattice is based on four (4) attributes: $\{A, B, C, D\}$ and that on running our minimality test on attribute-combinations of size 2 in the lattice, we discover that BC and CD are vulnerable to inference. In this case, our set of minimal augmented quasi-identifiers becomes: $\{BC, CD\}$. In the next step (i.e. considering attribute-combinations of size 3) we first determine if any of the combinations is a superset of either BC or CD. BC is a subset of ABC, BCD and CD is a subset of ADC, BCD. As a second step, we evaluate the remaining attribute-combinations. In this case, only ABD remains to be considered. ABD is considered to have an entropy that is lower than that of BC and CD so it is ignored. (If it were equal to or higher, it would have been included in the set of minimal augmented quasi-identifiers). Finally, we consider the combination $ABCD$. Since this has already been found to be a superset of BC and CD, it is not included in the set of minimal augmented quasi-identifiers. As such our set of minimum augmented quasi-identifiers is: $\{BC, CD\}$ since both elements allow for the discovery of all supersets of augmented quasi-identifiers that can be exploited for inference (re-identification).

4.7 Forming Safe Feature Sets of Attribute Combinations

Using the set of discovered quasi-identifiers, we must now outline a method of minimising information loss (due to suppressing the quasi-identifiers) and minimising homogeneity. We address this with a technique termed **compartmentation**. Compartmentation aims to create feature sets, based on attribute combinations drawn from the original dataset, by ensuring that the set of attributes contained in a feature set does not contain a subset of attributes that qualify as either a quasi-identifier or an augmented quasi-identifier. In line with this objective, we re-define anonymity as follows:

Definition 4.7 (Feature Set Anonymity) Let T be a subset of tuples within an un-anonymised dataset T_{UA}, and Q a combination of attributes such that a function $Q : T \rightarrow V$ maps the set of tuples $T = \{t_1, \ldots, t_m\}$ to a set V of all the possible values of the attributes within Q. A snapshot S of T_{UA} embodies feature set anonymity if and only if $\nexists S = \{t_1, \ldots, t_j\} \ni S \rightarrow T_{UA}$. That is, none of the records in S uniquely identifies any single record in T_{UA}.

We achieve feature set anonymity, by mapping the minimal augmented quasi-identifiers to an attribute graph denoted $G_A : (V, E)$ where V denotes the attributes (nodes) and E denotes the attribute-combination (edge). Each edge is weighted by the cardinality (or entropy) associated to the attribute-to-attribute combination. For example, $B \longleftrightarrow C$ represents the attribute combination BC, and $B \longleftrightarrow (10)C$ indicates that the degree of uniqueness in attribute-combination (BC) is 10. A high uniqueness value, evaluated based on a pre-defined threshold, indicates that the attribute combination under consideration poses a high disclosure risk. Alternative methods of evaluating per attribute combination disclosure risk include probabilistic approaches based on Bayesian networks which we studied in [173, 175]. Other works with a similar philosophy but drawn from the security field, include: [105, 154].

As a first step, we aim to reduce the disclosure risk by applying standard transformation measures such as *local suppression*, *generalisation* and *perturbation* to reduce the cardinality (uniqueness) manifested by the attribute combinations [213]. If the level of information loss does not affect the quality of the data, in that data analytics results remain accurate without impacting negatively on privacy, we stop at this point and the dataset can be shared.

Table 4.3 Base dataset for feature set creation

Name	ID-Number	Age	Gender	Postcode	Disease	EyeColour
Alice	1827–88	34	Female	34521	COVID-19	Blue
Bob	1727–88	32	Male	34530	COVID-19	Brown
John	1526–98	33	Male	34540	COVID-19	Black
Jeremy	5555–88	37	Female	34551	Lung cancer	Green
Jamie	6654–78	38	Female	34570	Gallstones	Blue
James	2345–66	36	Female	34580	Malaria	Blue
John	7896-54	42	Male	34555	Thyphoid	Blue
Jamie	6789-23	43	Male	34560	Lung cancer	Green
James	8967-23	44	Male	34525	Stomach cancer	Green

Table 4.4 Base dataset for feature set creation

Age	Gender	City	Disease	EyeColour
38	Female	Berlin	Lung cancer	Green
38	Female	Berlin	Gallstones	Blue
38	Female	Berlin	Malaria	Blue
42	Male	Frankfurt	Thyphoid	Blue
42	Male	Frankfurt	Lung cancer	Green
42	Male	Frankfurt	Stomach cancer	Green

Otherwise, in the second step we consider measures to break the functional dependencies that enable inferences (disclosures). To achieve this, we compute feature sets of G_A to determine which attribute combinations can co-exist safely (in a privacy-preserving way). This involves evaluating attribute combinations which are not included in the set of augmented quasi-identifiers.

Finally, as a post-processing measure, all feature sets are tested (using the cardinality and entropy metrics—see Sect. 4.3) to ensure no inference causing attribute-combinations exist therein (i.e. the created feature sets are free of quasi-identifiers).

Example 4.3 (Creating Feature Sets) Consider for instance, Table 4.3 in which we have two explicit identifiers say *Name* and *ID-Number*.

Preprocessing, results in the deletion of both attributes to form Table 4.4.

Suppose that after an analysis for augmented quasi-identifiers, we discover that there exist two, namely: $City, EyeColour$ and $City, Disease$. To anonymise the dataset we begin by creating feature sets centered around the identified quasi-identifiers. In this case, we obtain two feature sets as follows (Table 4.5):

So we create feature sets that do not place $City$, $Eyecolour$, and $Disease$ in the same dataset (Table 4.6).

We now discuss some empirical results aimed at testing the performance and accuracy of the scheme we have presented.

Table 4.5 Feature set based on the quasi-identifier *City*, *EyeColour*

Age	Gender	Disease	EyeColour
38	Female	Lung cancer	Green
38	Female	Gallstones	Blue
38	Female	Malaria	Blue
42	Male	Thyphoid	Blue
42	Male	Lung cancer	Green
42	Male	Stomach cancer	Green

Table 4.6 Feature set based on the quasi-identifier *City*, *Disease*

Age	Gender	City
38	Female	Berlin
38	Female	Berlin
38	Female	Berlin
42	Male	Frankfurt
42	Male	Frankfurt
42	Male	Frankfurt

4.8 Empirical Results

In order to conduct the empirical analysis, we focus on quasi-identifier discovery since this is important in determining where the risk of inference lies, and we also study information loss as a means of assessing the useful of the resulting datasets. We did not conduct a direct comparison to similar syntactic anonymisation schemes (see discussion in Chaps. 2 and 3) because the approach we have presented operates quite differently.

We recall that quasi-identifier discovery is in $O(2^n)$ time in the number of attribute combinations evaluated privacy disclosure risk [172, 175]. As mentioned before, we employ the Pascal triangle heuristic to select a minimal set of attribute combination candidates to evaluate. This reduces the runtime of quasi-identifier discovery to $O(2^{n-1})$ which in practical terms reduces the runtime by about a half (see Sect. 4.3).

In the next subsection, we present our empirical environment and the datasets used.

4.8.1 Experimental Setup

As mentioned in Chap. 3 (see Fig. 3.1), the goal is for the quasi-identifier discovery scheme to support the anonymisation service provider. Discovering quasi-identifiers in large high-dimensional datasets that are formed from compositions of data originating from different data owners, is a performance and data-intensive process. As such, for the experiments we employed a hardware platform aimed at enabling such large-scale processing operations efficiently. Furthermore, to access the impact of handling various datasets we consider ones originating from domains that provide rich sources of high-dimensional data but that also contain a significant proportion of explicit personal identifying data. Using such datasets is also useful in evaluating how well our proposed approach performs in terms of preventing privacy violations.

4.8.1.1 Hardware

We conducted our experiments on dedicated hardware, that hosted an AMD EPYC 7702 server comprised of 64 CPU cores, a maximum boost clock of 3.35 GHz, and 256MB of L3 cache. In addition, our server was endowed with random access memory capabilities of 64GB DDR4 ECC and supported with a 3200MHz processor, as well as storage of size 2TB SSD RAID 10 and an and an Nvidia Tesla V100 GPU accelerator.

4.8.1.2 Datasets

We evaluated our proposed approach with three datasets, namely Twitter, CrossFit Fitness, and Health datasets. The Twitter and CrossFit datasets are composed of real-world data, whereas the health dataset is a synthetic dataset generated on the basis of real-world attributes and attribute values. The Twitter dataset contains 500,000 records and 31 attributes, while the CrossFit dataset contains 1,572,222 records and 33 attributes. In the health dataset, we have 1,000,000 attributes and 101 records. A summary of the datasets and their characteristics is provided in Table 4.7.

An important characteristic of these datasets is that they are high-dimensional and large. As such processing them efficiently with existing syntactic data anonymisation algorithms (e.g. k-anonymity, l-diversity, ...) is infeasible [10].

4.8.2 Quasi Identifier Discovery

We evaluate the cost of augmented quasi-identifier discovery with respect to the datasets. For variability and to assess performance, the experiments were conducted with a different number of tuples and attribute combinations, and averaged for all three (3) datasets for comparison.

Table 4.7 Empirical datasets used

Dataset	Number of records	Number of attributes
Twitter	500,000	31
CrossFit	1,572,222	33
Health	1,000,000	101

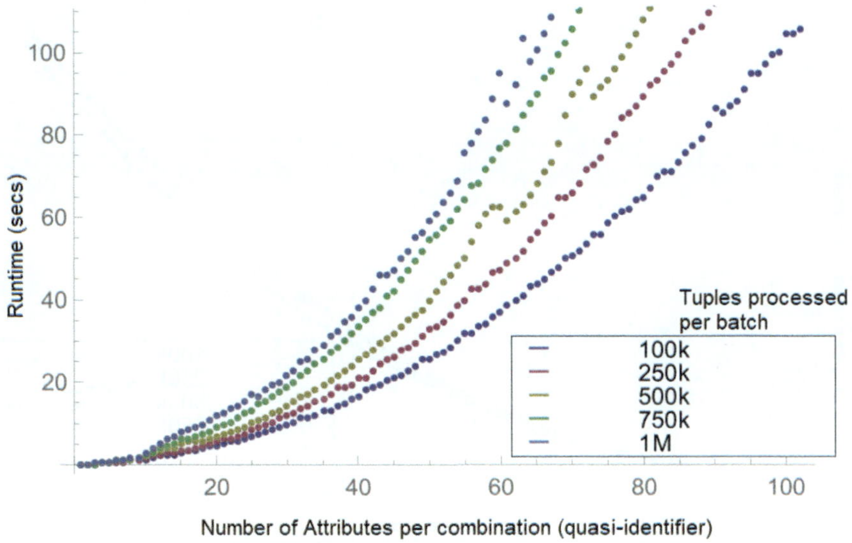

Fig. 4.2 Runtime performance of quasi-identifier discovery per batch sizes of tuples evaluated

As shown in Fig. 4.2 the runtime performance of the quasi-identifier discovery algorithm is evaluated for batches of tuples of size ranging from 100,000–1,000,000 over quasi-identifiers of size 10–100 in line with the datasets under consideration. The results indicate that while the number of tuples does not influence the runtime performance of the quasi-identifier discovery process, the number of attributes in a quasi-identifier does significantly. This result is corroborated by the result shown in Fig. 4.3, the runtime performance of quasi-identifier discovery stays relatively linear with respect to the number and size (number of attributes) of quasi-identifiers evaluated. This result aligns itself with the theoretical observation, in terms of the performance expectations of cardinality and entropy computations.

As a next experiment, we evaluated the vulnerability of the anonymised datasets to inference, on all three (3) datasets. In order to do so, we evaluate the information gain on a per compartment basis, and solely with the anonymised data. No auxiliary data is used. To do this, we use the information gain metric presented in Sect. 4.3. We compare the information gain on both the anonymised and un-anonymised versions of all three datasets. As no quasi-identifiers remain for any given feature set, we conclude that the anonymised dataset has a very low risk of vulnerability to cross-linkage (see Table 4.8).

As a further step, we increased the sizes of the feature sets to determine if this affects information loss and by extension the reliability of data analytics results. As shown in Table 4.9, information loss remains unaffected because by generating feature sets we minimise information loss. Furthermore, since we maintain the underlying data distributions data analytics results remain unaffected. Therefore the transformations that engender the feature sets do not provide any information that

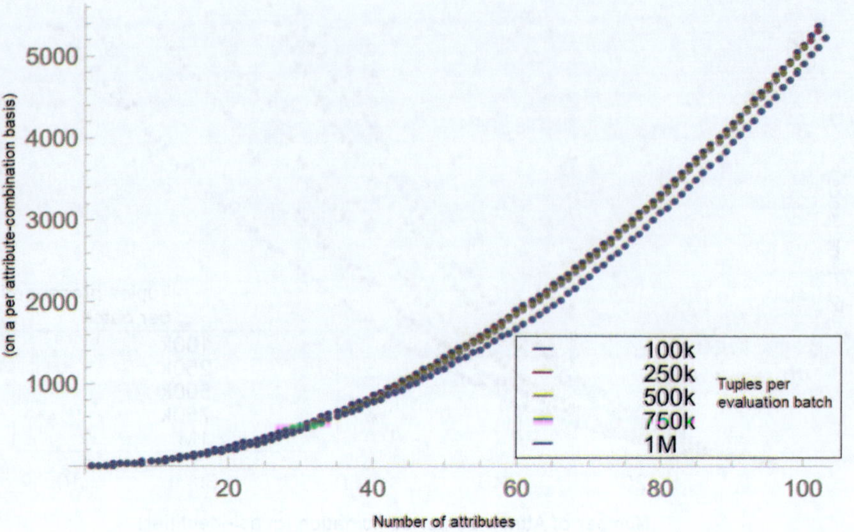

Fig. 4.3 Runtime performance of quasi-identifier discovery per attribute-combinations evaluated

Table 4.8 Feature set example where disease and SNP represent a quasi-identifier [172]

City	Disease	SNP	..
..	..	..	
Hamburg	DOID:3393	rs104894190	..
Hamburg	DOID:3393	*	..
Hamburg	DOID:1024	rs104894502	..
..	..	..	
=⇓	=⇓	=⇓	=⇓
City	Disease	SNP	
..	..	..	
Hamburg	DOID:3393	NULL	..
Hamburg	NULL	rs104894502	..
Hamburg	NULL	rs104894190	..
Hamburg	DOID:1024	NULL	..
..	..	..	

could support private information disclosure, but at the same time result in datasets that are useful for the data analytics operation.

We also evaluated the vulnerability to inference based on the semantics of the anonymised data and auxiliary data. In this case, we used the Twitter dataset as a case study and drew auxiliary data from the health and Crossfit datasets. Given the same reasoning, no unique attribute combinations (quasi-identifiers) were discovered and therefore no common base for cross-reference was established.

Table 4.9 Logistic regression coefficients as scaled weights for attributes [172]

Attribute	Original coefficients	Composition a coefficients	Compartmentation coefficients	Perturbation coefficients	Suppression coefficients
Age	100.00	100	32.9	0	0.05
Centimeters	49.4	37.5	100	100	6.8
drug_0	63.4	0	4.3	0	100
BloodType	33.9	8.8	45.1	0	0.05
Kilograms	50.5	62.3	24.3	0	0
snp_0	0.9	1.2	1.4	0	0
drug_1	2.1	2.3	2.8	0	0
drug_2	0.04	0.09	0.1	0	0

Table 4.10 Empirical characteristics of datasets

Datasets	Records	Attributes	2nd-QID	1st-IDs
Twitter	500,000	31	205	4
CrossFit	1,572,222	33	497	1
Health	1,000,000	101	5216	6

4.8.3 Performance on Various Datasets

To demonstrate that our proposed approach works for different datasets, in this experiment we set a time limit (TL) of 60 minutes and a memory limit of 32 GB, and we assess the number of augmented quasi-identifiers and explicit personal identifiers discovered. Table 4.10 summarises the runtime of our approach on the different datasets.

In order to evaluate information loss in terms of the ratio of transformations we made to the feature sets to prevent disclosure, we quantify the information loss in terms of the number of explicit personal identifiers eliminated from the dataset and the amount of processing required to transform augmented quasi-identifiers.

To compute the information loss in terms of explicit personal identifiers, we basically compute the number of attributes eliminated with respect to the size of the dataset, based on the cardinality metric and entropy. For augmented quasi-identifiers, we use the following formula [45]:

$$Loss = \frac{d}{n} + \sum_{i=1}^{q} \frac{g_{A_i}}{h_{A_i}} \times \frac{1}{q} \times \frac{n-d}{n} \tag{4.7}$$

where n is the number of attributes, d is the number of deleted/suppressed attribute values, q is the number of augmented quasi-identifiers, g_{a_i} is the generalisation level of attribute A_i in the generalised table, and h_{A_i} is the height of the generalisation hierarchy of A_i.

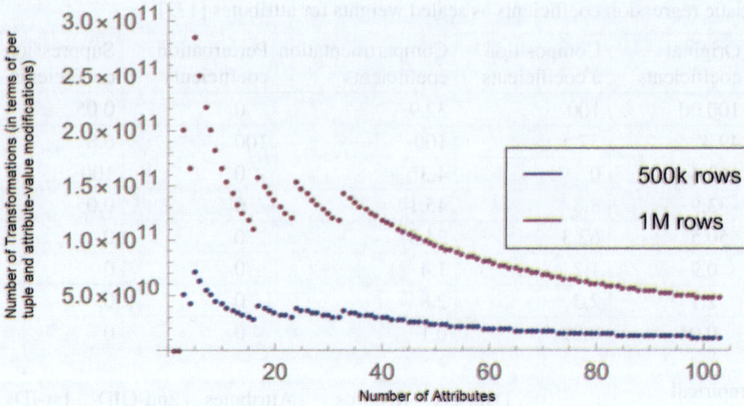

Fig. 4.4 Growth of information loss with respect to growing quasi-identifier sizes

This formula enables us to evaluate the impact of generalisation and suppression (row-wise or per attribute value-wise) on information loss. Since feature set formation is supported by global recoding, all values of an attribute are generalised to the same hierarchy level value. Information loss due to generalisation is evaluated by $(\sum_{i=1}^{q} \frac{g_{A_i}}{h_{A_i}})$ multiplied by the remaining records to which generalisation is applied $(\frac{n-d}{n})$ while that due to suppression is evaluated as a percentage of eliminated rows $(\frac{d}{n})$.

Figure 4.4 shows the evolution of information loss over increasing columns and therefore growing describing data points per records. We note that the decreasing levels of information loss relative to the size of the quasi-identifier is due to the fact that the anonymisation mechanisms operates on a per-row basis. This is advantageous in that alterations (transformations) to attribute-values are only conducted if the attribute-value raises the disclosure risk and in turn, reduces the rate of information loss.

4.8.3.1 Query Accuracy (Data Usability)

Since information loss has an impact on the veracity of the data, we consider the usability of the data as an indicator of whether our proposed approach generates data that is both privacy-preserving and usable in terms of providing accurate query responses. In order to do this, we ran several queries against the original (un-anonymised) dataset and the anonymised datasets and compared the results in terms of similarity using the Euclidean distance.

Figure 4.5 depicts the query accuracy over growing sizes of quasi-identifiers. As observed before (see Fig. 4.5), we note a similar decrease in information loss and correspondingly, an increase in query accuracy. This is expected since richer feature sets characterised by more attributes implies that more information is available,

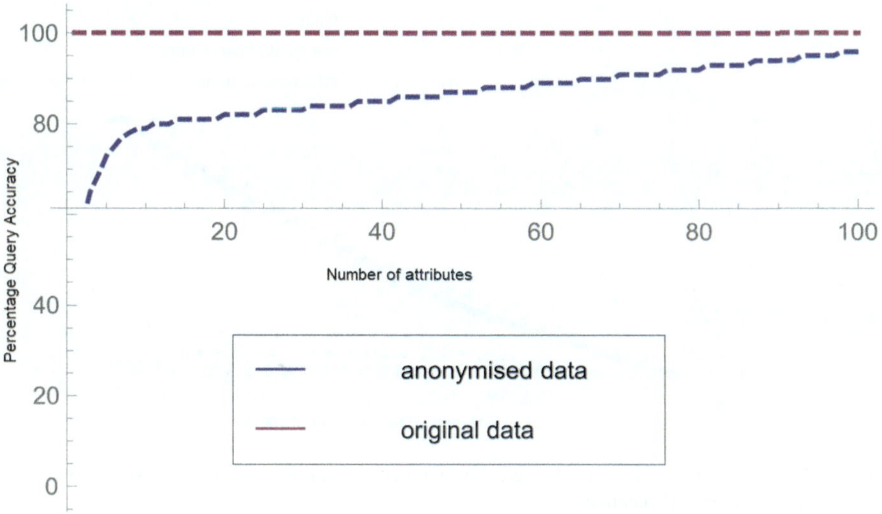

Fig. 4.5 Growth of percentage query accuracy with respect to growing quasi-identifier sizes (comparing accuracy on anonymised data relative to original data)

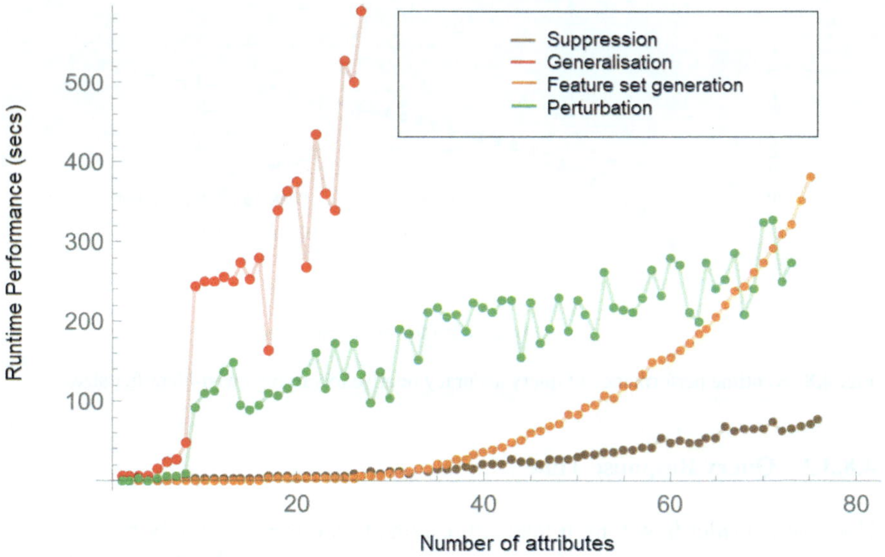

Fig. 4.6 Runtime comparison vis-a-vis state-of-the-art syntactic anonymisation schemes

thereby positively impacting information loss and query accuracy. Furthermore, as Fig. 4.6 illustrates, feature set formations scale well in comparison to standard data transformation mechanisms in terms of runtime performance for generating privacy preserving datasets (Fig. 4.7).

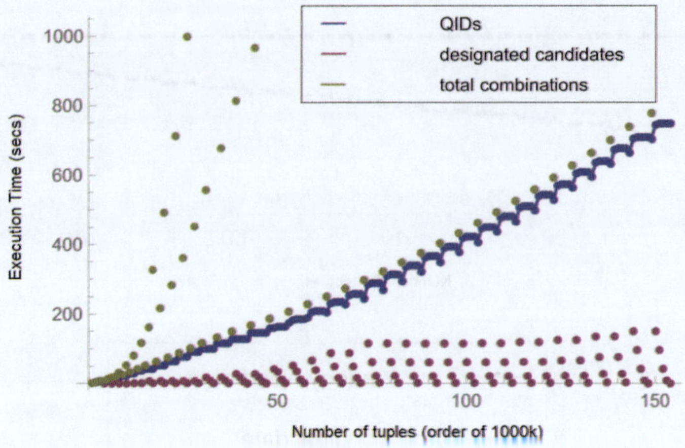

Fig. 4.7 Runtime performance relative to growing quasi-identifier sizes on query accuracy with respect to varying tuple batch sizes

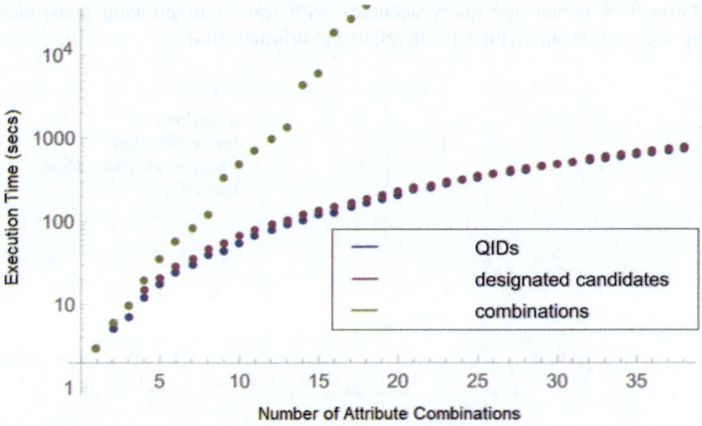

Fig. 4.8 Runtime performance of query accuracy relative to varying quasi-identifier sizes

4.8.3.2 Query Response Time

Here we consider how long it takes to respond to a query accurately by comparing the results over anonymised and un-anonymise datasets. In this way, we can determine whether or not the content of augmented quasi-identifiers affects query response time in terms of returning accurate results.

Since the transformation measures are applied a-priori to the data, the actual query response time is unaffected. Figure 4.8 shows in spite of the increasing sizes of the quasi-identifiers by using a greedy heuristic to select the starting point for evaluating quasi-identifier candidates (see Sects. 4.5 and 4.6), we can control the processing overhead incurred to ensure that the runtime performance plateaus with increasing quasi-identifier sizes. This is also emphasised in Fig. 4.7, where fewer quasi-identifier candidates need to be processed to find the minimal augmented quasi-identifier.

4.9 Discussions

In this chapter we presented a method of anonymising high dimensional data with low information loss. Our proposed approach operates by discovering all the existing quasi-identifiers in a given high-dimensional dataset, and by creating feature sets of data by breaking the functional dependencies that exist between the quasi-identifier attributes. Creating such feature sets based on attributes that can be published together safely, helps avoid having to apply excessive generalisation, suppression or perturbation when syntactic anonymisation mechanisms to the data, and as such reduces information loss considerably without sacrificing privacy. An added benefit is that since our proposed approach to data anonymisation is supported by an efficient quasi-identifier discovery algorithm, it scales well to larger datasets. By discovering quasi-identifiers efficiently, attacks that exploit low information loss on anonymised datasets (e.g. minimality and semantics-based attacks) can be circumvented.

We compared the performance and privacy of our algorithm on three (3) different datasets that while being large and high-dimensional, contain different types of high dimensional data namely, a Health dataset that contains a mix of textual and numerical data, as well as a high number of attributes that qualify as 1st class-identifiers; a Twitter dataset that contains a high degree of textual and auxiliary data that increases the re-identification probability when standard transformation mechanisms are applied; and a CrossFit dataset that has a mix of similarities to the Health and Twitter datasets. The idea was to test our proposed solution on different datasets to determine its effectiveness.

Our results indicate that the proposed algorithm processes underlying data highly efficiently, and successfully finds all quasi-identifiers of any length (size in terms of number of attributes) in a given dataset. Without these unique attribute values, provoking privacy disclosures based on direct linkage is challenging and as such provides a reliable mechanism for protecting against various re-identification attacks.

We note however, that quasi-identifier discovery is a performance intensive process and have as such conducted work to propose approaches to enable more efficient quasi-identifier discovery in large high-dimensional datasets. Since this falls beyond the scope of this monograph, we refer the reader to our work on Bayesian networks and parallel quasi-identifier discovery with GPU processing [175, 176].

A further benefit of the quasi-identifier discovery we have discussed is that it can be employed as a verification mechanism to determine privacy disclosure risks in already anonymised datasets. This can be used as a privacy rating metric to support assertions of privacy in datasets and offers the benefit of providing the data analytics mechanisms with a means of pre-processing datasets to ensure that they are privacy preserving before conducting analysis on the data. Other potential benefits lie in alleviating issues such as base rate neglect which can arise due to skewed distributions of anonymised data [174].

In the next Chapter, we discuss mechanisms for enabling expressions of user privacy preferences and how these support our premise of privacy as a subjective notion. We consider the implications of user privacy preferences on generating safe anonymised datasets, and the impact this has on composing privacy-preserving datasets based on data originating from multiple individual data owners.

Chapter 5
Accounting for User Privacy Preferences

In Chap. 3 we presented a mechanism for dealing with anonymising lossy data with low information loss, and without putting privacy at risk. Chap. 4 built on the discussion in Chap. 3, to look in depth at the issue of efficient quasi-identifier discovery as a means of guaranteeing privacy. As we mentioned in Chap. 2, quasi-identifier discovery is central to enabling syntactic data anonymisation and is particularly relevant in scenarios that involve compositions of large datasets emanating from multiple (independent) distributed data owners.

A further consideration, and one that we would like to focus on in this chapter, is the issue of privacy preferences, and how varying privacy preferences must be accounted for in guaranteeing collective privacy. In Chap. 1, we raised the issue of subjective privacy and pointed out that while subjective privacy might be acceptable in everyday life, in the digital sphere, individual privacy preferences affect (positively and/or negatively) collective privacy and so must be accounted for in privacy assertions. It is important to note that while we will discuss this primarily with respect to data anonymisation, expressions of individual privacy preferences emerge in several web applications (e.g. social media, which we discuss in Chap. 6) and so it makes sense to give some consideration to its implications on data anonymisation operations.

As a basis for accounting for user privacy preferences in generating privacy preserving datasets, in this Chapter, we present and discuss how user privacy preferences can be captured and translated into anonymisation objectives. In order to better understand user privacy preferences we conducted a proof-of-concept survey to understand what impressions people have about the protection of their data and whether or not there is a general understanding of the implications of sharing data with third party applications.

One method of translating the survey results of user privacy preferences is to use multinomial regression and association rules to determine what factors or attributes influence user privacy preferences and translate these into a privacy barometer [186, 189]. As discussed in depth by Sakpere et al. [186, 189, 190], using multinomial

A. V. D. M. Kayem, *Enabling Privacy Preserving Data Analytics*, Advances in Information Security 92, https://doi.org/10.1007/978-3-031-93906-8_5

regression and association rules is also useful for relatively small datasets, in that both techniques offer useful insights on the relationships between variables [102].

Using such insights regarding user privacy preferences can be used to support the design of a three tiered privacy policy and eventually anonymisation (e.g. k-anonymisation) [186, 189] but also in terms of handling conflicting privacy decisions [152]. We discuss the approach based three tier privacy as a basic method of dealing with varying privacy preferences using a classification metric.

5.1 Privacy Preferences

The anonymisation techniques that we have discussed so far, use the same privacy levels for all the user records in the dataset [192, 213]. For instance, with k-anonymity models, for very high privacy the aim would be to have very large sized equivalence classes and vice versa for low privacy requirement settings. However, in certain applications, using the same privacy level for all the users in the dataset is not practical feasible. Two (2) reasons for this reasons are: first, privacy is subjective and so, people tend to have varying opinions of privacy protection requirements for their data [89, 236]; and second, high privacy requirements can result in unnecessary information loss rendering the resulting data useless for analystics purposes. Furthermore, mis-representations of individual user (data owner) privacy needs make it harder to anonymise datasets for scenarios in which sub-optimal privacy might suffice [35, 145, 226, 236].

An example arises in the following scenario where "Alice", whose car was stolen, does not want her insurance company to learn about the incident before she has had a chance to inform them personally. In this case, to account for Alice's personal privacy preferences, one approach would be to use a mechanism such as MOA (see Chap. 3), with a sensitive attribute value ("Car Theft") with a high-severity weighting based on Alice's privacy preference. On the other hand, "Bob", whose mobile phone was stolen, might be less concerned about letting his insurance company know about this, and might not mind having his true identity revealed. In this case, again using the MOA approach this would translate to placing a lower severity weighting on the sensitive attribute-value associated with Bob (e.g. "Phone Theft"). If we had applied the standard MOA approach, depending on the type of phone and/or car we would have placed a high severity weighting on Bob's record and fully anonymised both records thereby incurring higher levels of information loss than would have been the case with the alternative. Based on real-life interviews with crime victims, one observation we made was is that individuals perceive crime severity as a function of whether or not being associated with the crime stigmatises either the victim or the perpetrator [188].

However, while anonymsing data to account for user preferences can offer a good method of reducing information loss in anonymised datasets, without putting individual privacy at risk, it is important to also have computational approaches for handling biases, ensuring fairness, and auditing privacy decisions. Furthermore, the implications of users' (participants') individual privacy choices on the collective privacy. In this Chapter we look at another approach, in addition to the MOA approach discussed in Chap. 3, to accounting for individual privacy and factoring this into the generation of anonymised datasets that account for collective privacy.

5.2 Three-Tiered Privacy

In order to account for user privacy preferences, one approach is to employ a three-tiered privacy scheme in which personalised privacy is labelled according to three different levels of privacy, namely: "Low", "Medium" (neutral), and "High". The basis for choosing a three-tiered personalised privacy scheme is that it is flexible and simple without shifting the onus of the anonymisation scheme from its overarching objectives [44, 125, 241]. However, a more sophisticated approach would be to use some form of optimisation in which various privacy metrics are weighted. For instance, we could factor in metrics such as bias, fairness, behavioural factors, and assess privacy based on pre-defined thresholds for each parameter [9].

From the usability perspective, a three-tiered privacy scheme also simplifies the process of assigning a user's privacy preferences to a privacy level and accounts for the degree of sensitivity of the data that needs to be anonymised. In other words, for data with a high proportion of highly sensitive attributes, the scheme would opt for the maximum privacy setting for each choice, and would also employ a metric such as the severity weighting which we discussed in Chap. 3.

5.2.1 Low Privacy

At this level the focus is on data that has a low number of sensitive attributes. For instance, referring back to Chapter3 the severity of the crime of rape differs from that of the crime of theft. So in this case we could say that "theft" is ranked lower in terms of severity to privacy exposure "Hacking government web services". Therefore as we explained in Chap. 3, the scheme needs to account for sensitivity of sensitive attribute values. To associate the severity weighting to user privacy preferences, in the case of users who opt for no anonymity, the privacy requirement at this level represents users who have no objection to their identity being revealed. This implies that data provided by such users can be used to support high data utility operations. This translates into removing the sensitive attribute(s) severity weighting requirement for such users when creating equivalence classes based on considerations for distributions of sensitive attributes. We note here that in this case the dataset would be separated from other datasets containing information about users with stronger privacy requirements. This procedure would follow an approach similar to the feature set generation approach that we discussed in Chap. 4.

5.2.2 Medium Privacy

At this level, the focus is on data that cannot be classified as having either a highly sensitive nature or a low sensitive nature. This level also represents users who want their data protected in such a way that information released to third-party service providers retains a high utility value, but at the same time protects critical aspects of their personal details. We assume that users at this level want an equal balance between data utility and privacy enforcement.

5.2.3 High Privacy

The high privacy level focuses on data that is highly sensitive in nature. This privacy level represents users who care a lot about their privacy. Users at this level prefer maximum protection of their data. We can assume that users at this level prefer high privacy enforcement to high data utility.

Based on these privacy levels, in the following section, we discuss results of a proof-of-concept survey that we conducted to determine if such a three-tiered user privacy model is feasible in practice and an approach that users can relate to.

We quantify privacy levels using the following function:

$$P_L = \begin{cases} k & \text{if } L \geq 2 \\ k/2 & \text{if } L = 1 \\ 1 & \text{otherwise} \end{cases}$$

where P_L denotes the privacy preference level L such that $0 \leq L \leq 2$, and $k \forall 1 \leq k \leq max$ indicates the size of the associated equivalence class for clustering data points based on privacy level similarity. max denotes the largest possible value k can take on, but is constrained by the number of data points in the dataset. In this case, the highest privacy level is at $L \geq 2$, medium privacy level at $L = 1$ and the lowest privacy level at $L \leq 1$. We now discuss how privacy preference expressions are collected and then eventually translated into privacy levels.

5.3 Privacy Preference Expressions

To quantify privacy preferences, we surveyed participants aged between 21 and 40 years old who had been victims of different crimes and who regularly use various digital devices such as a smartphone. We assumed that the willingness of an individual to share information is inversely proportional to a user's personal privacy preference. Therefore, the survey aims to examine the privacy preference levels (i.e. low, neutral or high) of its subjects, and based on the data collected map this information on to either one of the privacy preferences we have discussed. The justification for choosing a three-tier personalised privacy setting is that according to Chuang et al. [44] and Yuan [241], users have varying levels of privacy and can realistically choose between three levels of privacy at most [147]. The survey also collected information on variables believed to motivate these choices [16, 100, 121, 202]. Further details on the survey structure and questions, can be accessed in [185, 187].

We analysed the data with both visual and quantitative methods to summarise the contents of data collected to obtain preliminary information about the relationship the study participants and the 9 variables studied, namely [189]:

- Sex
- Age group (Age)
- Present Education Level/Occupation (PEL)
- Highest Education Qualification (HEQ)
- Victim of Crime (VoC)
- Crime Experienced (CEx)
- Preferred Privacy Level (PPL)
- Share with Third Party (STP)
- Reason for Choice of Privacy (RCP)

All the subjects interviewed were postgraduate students whose occupation and HEQ are exactly related. That is, for example, a student enrolled for a PhD has a Master's degree as his or her HEQ. As PEL gives some clue about HEQ. Similarly, the variable, STP, is closely related to the variable, PPL. That is, a user's privacy preference indicates his/her willingness to share data with third party. Thus without loss of generality, HEQ and STP are not included in Table 5.1 and Fig. 5.1 which provide summaries of the major categories of each of the variables.

All the 24 respondents reported their PEL as either PhD or MSc, except for respondent 19 who reported Mtech. Consequently, there were two categories of PEL, namely: (1) Master's; and (2) Doctoral. Similarly, the HEQ variable was re-classified as Bachelors, Honours and Master's.

Three (3) respondents did not provide information about the type of crime they had been a victim of, while the responses of the remaining 20 subjects varied among robbery, assault, burglary, car snatching, fraud, mugging, none and theft. These categories of the VoC variable were reduced to four namely; robbery, theft, other and none. It is important to clarify the contextual implication of some the related crimes such as *robbery, burglary* and *theft. Theft* is the mildest and basic of the trio which often takes place to deprive the owner of his or her possession. *Robbery* on the other hand makes use of coercion to take possession belonging to another person and is usually associated with blackmail, for instance. Finally, *burglary* implies gaining access illegally into a building to commit a crime such as theft.

Table 5.1 Categories of the surveyed variables [107, 108, 189]

Variable	Categories	Subjects
Sex	Male	18
	Female	6
Age group	25–30	11
	31–35	4
	36–40	9
Occupation	PhD student	17
	Masters student	7
Reason for privacy choice	None	5
	Aaptive	1
	Personal	8
	Insensitive	4
	Reduce crime	6
	Explicit ID delete	2
Crime victim?	Yes	20
	No	4
Crime experienced	Burglary	3
	Fraud	1
	Robbery	2
	Theft	12
	Mugging	1
	Assault	1
	Car snatching	1
Preferred privacy level	Low	8
	Neutral	9
	High	7

Thereafter, only theft and robbery, were retained in the final classification. The STP variable has only four records. Among those records, two are "No", one is "Yes" and the rest of the answers are neither "Yes" nor "No" (e.g. maybe). We combined the responses and labelled them as "*answered*", while the remaining 20 subjects were assigned to the category "*unanswered*". There was great variation among the responses to the RCP variable. However, nine (9) subjects attached the word "personal" to their responses, while five (5) subjects provided no information at all. In this case, the categories adopted for the RCP variable for the multinomial regression analysis were *personal*, *other*, and *unspecified* [185, 187]. Other studies showing how user privacy preferences can be translated into anonymisation goals include ones centered around personality analysis [135, 152, 218].

Based on the survey results, participants were classified as shown in Table 5.2.

In the next section, we consider an example of how user privacy preferences can be mapped to k-anonymisation.

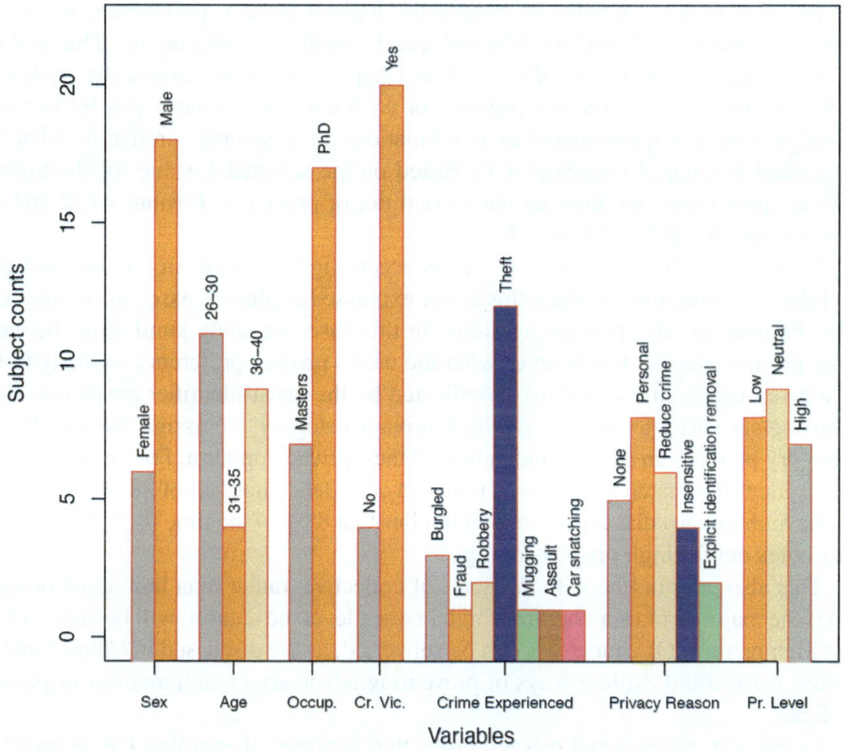

Fig. 5.1 Histogram illustrating the distribution of subjects over the different categories of variables surveyed in the primary study of privacy level preference [189]

Table 5.2 Participants and corresponding privacy opinion

Number of participants	Privacy opinion
10	Un-Declared
5	Neutral
9	Personal

5.4 Three-Tiered Privacy and k-Anonymity

In order, to match the three-tier privacy scheme to generated k-anonymised data, a threshold k-value is set to indicate the highest privacy level, and each of the three-tiers of privacy is then computed in reference to this. Depending on a user's assigned privacy level (see Sects. 5.2 and 5.3), which is determined based on the indicated privacy opinion, the k-values are used to determine the corresponding equivalence classes to which a user's data point will be assigned. As mentioned before, we set a value for k which serves as a threshold and guide for determining the values of k to assign to different privacy levels in the three-tiered hierarchy.

In order to set a k-value to handle the highest privacy preference, we assess various feature sets based on different quasi-identifier combinations. This is done in a manner similar to that discussed in Chap. 4, where we assess the trade-offs between information loss and privacy for each case. We achieve this by iterating through different quasi-identifier combinations in a manner similar to what we described in Chap. 4 (see Sect. 4.7). Based on the selected k-value for the highest privacy preference, we then set the second tier of privacy to k-value = $k/2$ and the lowest privacy tier to k-value = 1.

In order, to adjust the baseline values according to changes in the data and also to infer user preferences when this is not expressed explicitly, association rules can also be used to infer privacy opinions. In this case, we study similarities between user profiles and use this to infer what the user's privacy preference is likely to be. If a given user's characteristics as indicated by the quasi-identifier attributes, show a strong similarity to users in the high privacy category, we assign the user to this category pending an explicit indication of their privacy opinion. For example, if the quasi-identifier match for a user X falls in a confidence interval of 88% with respect to the high privacy classification, this implies that 88% of the time users with similar attributes opt for high privacy settings.

This also goes in line with the goals of collective, rather than individual privacy. So if the majority of users opt for a high privacy level the dataset, will be anonymised for high privacy. Of course this can be subverted, as we discussed in Chap. 2 and so future work could explore ways of preventing adversarial manipulations of privacy policies.

Essentially, three-tiered privacy protection is aimed at enabling the anonymisation mechanisms we have discussed in Chaps. 3 and 4, to account for user privacy preferences, in a bid to have more expressive representations of collective and individual digital privacy preferences. In the following Section we explain how privacy preferences are used in conjunction with sensitive attribute severity rankings to support minimising information loss while enforcing user privacy preferences.

We could also have alternatively used an approach similar to the one we discussed in Chap. 4, via a utility function, to determine the optimal privacy preference level to assign to a user based on his/her indicated opinion and the respective information loss. The aim is to account for both individual privacy preferences but to assess, based on the potential degree of information loss, what the impact on collective privacy will be.

In this case, a utility function could employ a weighted information loss metric, that is computed through an iterative function until a global optimum is found. At the jth iteration, the weighted information loss metric $LM_{k,j}$ for the k-Anonymisation algorithm is calculated based on the threshold k-value. Initially the assigned threshold k-value is the one that was pre-set to match the users' privacy preferences, however as the computation iterates to find a global optimum the k-value takes on the new computed value to become k_{new}-value. As is the case, in Chaps. 3 and 4, different attribute combinations for quasi-identifiers Q_T are evaluated to determine which ones result in the lowest information loss and highest privacy.

Table 5.3 Privacy preference weightings per crime case

Disease	Severity	Privacy preference
Theft	3	Medium
Murder	7	High
Lies	3	Low
Burglary	5	High
Robbery	5	High

Example 5.1 In Table 5.3, P_{L_a} denotes the list of privacy preferences with respect to selected privacy sensitive attributes and $P(\cdot)$ maps the privacy category to its weight, which in this case is simply the guideline sensitivity ranking for a given crime case (see Chap. 3 for sensitivity ranking). So, $P(\text{Theft}) = 5$ indicates a duration of 5 days. Following this scale, the risk of privacy loss for a tuple containing *"Murder"* is higher than for a tuple with *"Lies"*.

We now consider some empirical results and discuss their implications with respect to individual and collective privacy.

5.5 Empirical Analysis

We conducted experiments with the three-tier user-defined privacy preference model, based on the crime dataset used to support experiments in Chap. 3. We started off with a threshold k-value, progressively moving to a more personalized k-value. As a heuristic, the choice of a the initial threshold k value is guided by values used in published experimentation results [242], and from our results in Chaps. 3 and 4.

5.5.1 *Reduction of Excessive Privacy Control*

The results shown in Fig. 5.2, indicate that supporting k-anonymity with three-tiered privacy relaxes privacy constraints but does not impact negatively on the levels of privacy originally offered by the k-anonymity model. Our three-tier personalised approach has a 16.15% rate of excessive privacy control, while the Personalised model suggested by Gedik [89] and the non-personalised model incur information loss rates of 63.08 and 23.08% respectively.

One possible reason for this that the Gedik and the non-personalised privacy model apply a single k-value to the data, based on the overarching privacy goals. As expected, with the three-tier model higher k-values also incur high information loss rates on a per record basis.

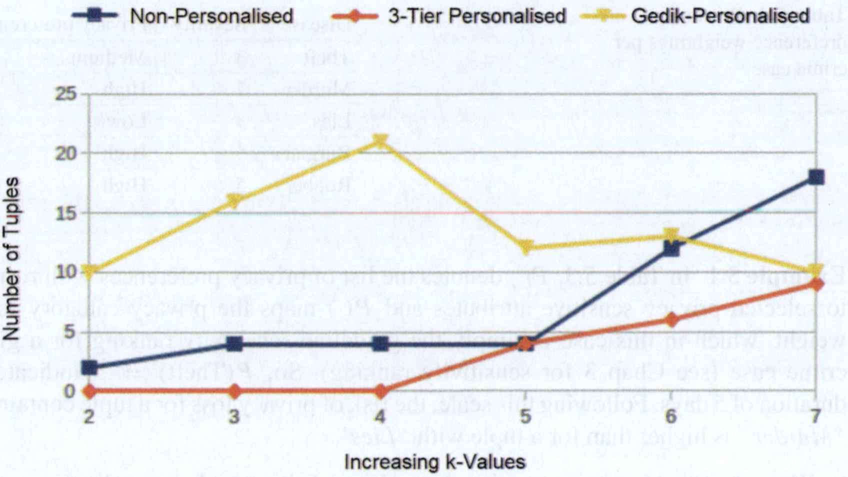

Fig. 5.2 Evaluating the effect of personalised, non-personalised, and Gedik privacy the data with respect to varying sizes of equivalence classes (size = k for $k \geq 2$ [189]

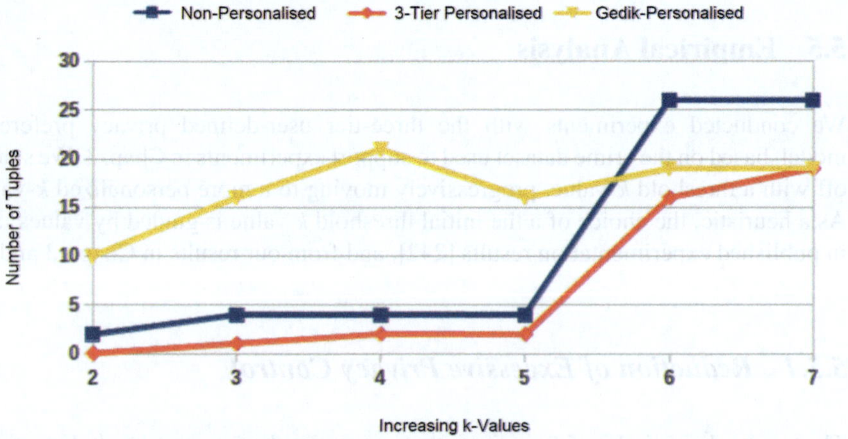

Fig. 5.3 Impact of the personalized and non-personalized privacy scheme on minimizing information loss due to suppressed records [189]

5.5.2 Record Suppression

As mentioned before, (see Chap. 2), using record suppression as a data transformation technique generally results in high information loss. In Fig. 5.3 the effect of personalized and non-personalized privacy on suppressed records is shown. In this case, using the three-tier personalised privacy model results in a lower rate of suppressed records, 26%, when compared to the Gedik personalized privacy model

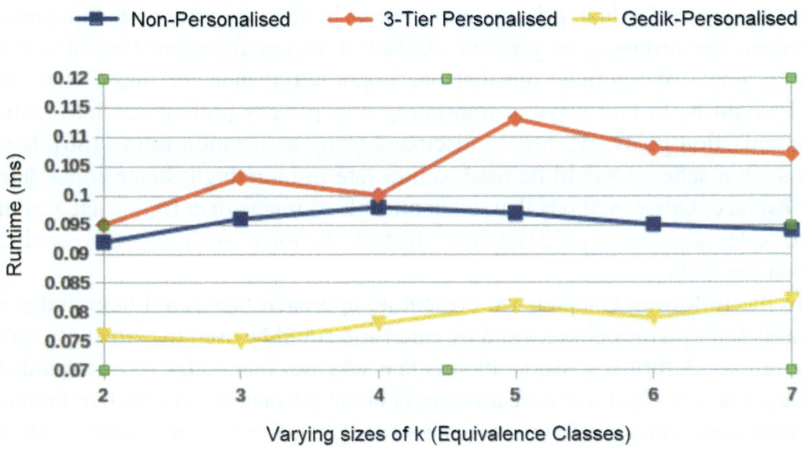

Fig. 5.4 Performance costs due to personalized and non-personalized privacy preferences [189]

that incurs a suppression rate of 77.7% and the non-personalized privacy model that incurs a 51% suppression rate.

This is due to the gains afforded by reducing the potential subset of high privacy records to roughly $\frac{1}{3}$ the size of the original dataset. Of course, in this case, it assumes that the dataset is equally segmented. In scenarios in which a high number of users opt fr the highest privacy settings, we are likely to see similar loss rates on the three-tier model, as with the Gedik and non-personalised schemes. As future work, one method of alleviating the information loss rates in this case would be to apply attribute compartmentation in the case of high privacy requirements.

5.5.3 Performance Costs

Figure 5.4 provides a performance comparison of the three-tier, the Gedik personalized and non-personalized privacy models.

We note that our approach has the highest computation cost since it first translates user privacy preferences into a corresponding k value representing a user's privacy preferences and then searches the dataset to form equivalence classes on the basis of the k-value representations.

5.6 Discussions

In this Chapter, we discussed a method of accounting for user privacy preferences in generating anonymised data. The motivation for doing so, was to consider a means of including individual privacy preferences in the objectives of anonymisation schemes, for instance, to generate datasets that are representative of collective privacy goals. We pointed out that one important avenue for future work in this field would be to find ways of translating user privacy preferences into realisable anonymisation goals. We briefly discussed using association rules [186], but note that such a scheme would be hard to translate to large high dimensional datasets, for instance. Other methods that are applicable to image data for instance, could be to consider personality [152] and even historical choices with respect to application use on the Web.

In the following Chapter, we present an approach that could help nudge users towards safer privacy choices and by extension could be used to extract user privacy preferences. A further consideration is that schemes that nudge users towards safer privacy choices can raise user awareness about privacy as a collective rather than an individual concept. However, since such schemes can also be subverted, future work into investigating ways of circumventing such activities is needed.

Part IV
Usable Privacy - A Discussion

Chapter 6
Privacy Recommender Systems

Recommender systems can offer an approach to alleviating privacy settings usability issues that users (data owners) face on shared content platforms such as social media. These social media platforms raise an example of shared data belonging to different data owners that oftentimes is composed in the form of joint data such as shared blogs, posted content, forums, or chatrooms. Oftentimes users in these contexts have varying privacy preferences, as discussed in Chap. 5, but are not always aware of the implications of their personal privacy preferences on collective privacy. Furthermore, matching usability with privacy provision is a challenging problem.

In this chapter, in line with our goals of enabling low powered and low processing privacy solutions, we present an example of a recommender system supported by the Random Forest Algorithm (RFA) algorithm, as a simple performance efficient approach to supporting users (data owners) in making privacy-aware decisions. As a case study, we consider how recommender systems can support users in configuring privacy settings to ensure that the data they share is protected. To this end we explore two approaches—the first is based solely on guiding the recommender system with user input, while the second approach employs privacy policy correlation graphs. The results indicate that the accuracy levels with both approaches is similar implying that automated privacy mechanisms can be designed to anticipate user privacy preferences and adjust collective privacy settings accordingly [3, 4].

6.1 Rationale

Popular SMPs such as Facebook, Twitter, and Instagram have an average of 1 billion active users the majority of whom access these platforms on a daily basis [182, 201]. While Social Media Platforms (SMPs) have become integral to the way in which users interact and communicate online, they also provide multiple ways in which

A. V. D. M. Kayem, *Enabling Privacy Preserving Data Analytics*, Advances in Information Security 92, https://doi.org/10.1007/978-3-031-93906-8_6

sensitive personal identifying information can be shared on sometimes publicly accessible areas [205]. For instance, roughly 350 million photos are shared on Facebook daily and 500 million tweets are sent on Twitter [79, 182, 201].

The ease with which such sensitive information can be accessed, and by a large number of people, puts SMPs users at risk of both privacy and security violations such as identity theft, financial fraud, cyberstalking, and cyberbullying [79, 87, 167]. To guard against this, most SMPs implement mechanisms to enable users configure privacy settings. Privacy settings are essentially a set of rules that enable users to control who can access a users' disclosed information. For example, SMPs like Google+ and Facebook have fine grained privacy settings that allow users to control access to individual profile attributes such as *birthdate*, and *address*, as well as to user generated content such as *posts, photos*, and *videos*. However, existing work indicates that users frequently fail to configure their privacy settings due either to lack of awareness, or the complexity and mental burden required configure the settings correctly [80, 130, 140, 149].

Various approaches have been proposed to assist users with privacy setting configuration [18, 77, 91, 194, 199, 201]. However, most are limited to either configuring privacy settings for profile attributes, or configuring privacy settings for user-generated content. Furthermore, accessing privacy settings on most existing platforms is complex and tracking the effect of changes is challenging for users. This is problematic, because both profile attributes and user-generated content can contain sensitive information. For instance, users have been documented to have put their real cell phone numbers on their profiles, as well as information about their location in status updates [90, 94, 95, 217]. Therefore, protecting profile attributes without also protecting user-generated content (and vice-versa), is not a good privacy protection solution for keeping personal information secret or protected.

We now present an automated privacy policy recommender system to address this problem [3, 4]. The recommender system consists of two main components namely, a *Profile Attributes Protector* (PAP), to handle recommendations for profile attributes; and a *User Content Protector* (UCP) for user-generated content. We discuss both approaches in the following Section.

6.2 User Privacy Settings Recommender

The privacy settings recommender system is designed as a server-side solution. In our setting, this could potentially be based for instance at the anonymisation service provider's end (see Fig. 3.1 but depending on the system's architecture, could also reside at the client-end. For simplicity however, the model we describe in this chapter assumes a server-side implementation.

The privacy settings recommender system (which we refer to henceforth, as *the recommender*) consists of two independent components, that work in parallel to protect sensitive information. The *Profile Attributes Protector* (PAP), uses privacy settings that existing experienced SMP users have specified for their profile

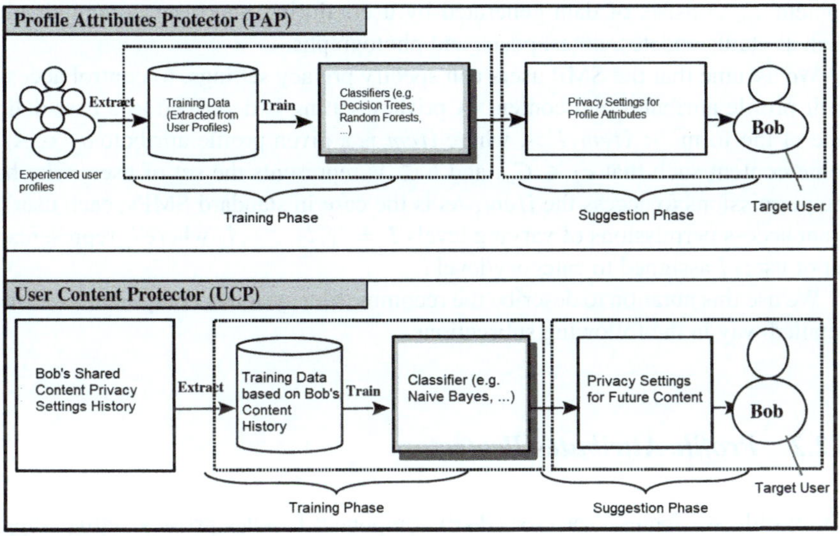

Fig. 6.1 An overview of the privacy policy recommender system [3]

attributes, to propose privacy setting configurations to target users, while the *User Content Protector* (UCP), uses the target user's privacy settings history, to suggest suitable privacy settings for the content that the target user might share in the future.

As depicted in Fig. 6.1 the PAP component extracts data from the profiles of experienced users, which is then used to train decision tree classifiers which output privacy settings suggestions to the users. The UCP component works similarly by extracting certain data from the target user's privacy policy history which is then used to train a Naive Bayes classifier which outputs privacy settings suggestions for user generated content.

Before we describe both the PAP and UCP components in more detail, it makes sense to provide a formal notation for the recommender system.

6.2.1 Notation

For simplicity, our SMP is represented using an undirected graph $G = (V, E)$, where V is the set of users who have a profile on the SMP, and E represents the set of social relationships (edges) between the users (nodes). Every user $u \in V$ creates and maintains a profile P_u, which serves as a digital representation of the user. Each P_u consists of a set of attributes $A_u = a_{u_1}, a_{u_2}, \ldots, a_{u_n}$ as well as a collection of user-generated content $C_u = c_{u_1}, c_{u_2}, \ldots, c_{u_k}$. The profile attributes A_u describe the profile owner (i.e. the user). These profile attributes include demographic information such as: *age, gender, political views/interests* . The user-generated content C_u consists of data generated by users during inter-user communications such as status updates, notes, posts, and photographs.

We assume that the SMP users can specify privacy settings, to control access to their profile attributes and content. A privacy setting is described as a user-defined rule of the form $< item, l >$, where $item$ is a given profile attribute $a_j \in A_u$ or some content such that $c_j \in C_u$, and $l \subset V$ represents the set of users who have the permissions to access the $item$. As is the case in standard SMPs, each user can grant access permissions of varying levels $L = l_1, l_2, \ldots, l_k$ where l_i represents the set of users l assigned to category/level i.

We use this notation to describe the recommender system's components in a more detailed way in the following subsections.

6.2.2 Profile Attributes Protector

The Profile Attributes Protector (PAP) component handles privacy setting suggestions targeted at all the profile attributes related to a user's profile. In order to do so, the PAP component employs three (3) steps to suggest privacy settings for a given profile attribute, say $a_i \in A$, as follows:

1. The PAP component starts off with a data collection phase during which data from profiles of existing SMP users. From every profile, we extract the user's demographic information, and the privacy settings associated with the attributes (a_j). The collected information is then stored in a dataset D, in which every tuple is represented as (F, l_j), where $F = \langle f_1, f_2, \ldots, f_m \rangle$ are the features representing the user's demographic attributes, and l_j is attribute a_j's privacy setting.
2. Next, we train a classifier to predict the privacy policies of the a_j attribute. As a first step, we used a decision tree model, but as we show later in the experiments this could be replaced by any other existing classifier model, based on performance constraints. We decided to use decision tree algorithms, because they are capable of handling classification of categorical, noisy, and incomplete data such as our training data D [228].
3. Finally, in the suggestion phase, when a target user registers on the SMP, we extract demographic information from his/her newly created profile, and pass it to classifier, which in-turn predicts a privacy setting for each of the attributes included in the demographic information (in this case the attribute a_j). This privacy policy is then simply suggested to the target user. Figure 6.2 illustrates the process of suggesting privacy settings for individual profile attributes, that we have described above.

So far we have discussed how the PAP component works to suggest privacy settings for a single profile attribute, say a_j. In order to suggest privacy settings for all the attributes in the target user's profile we build a series of training datasets $D_1, D_2, \ldots, D_{|A|}$, one for each profile attribute $a_j \in A$. Next, we use these

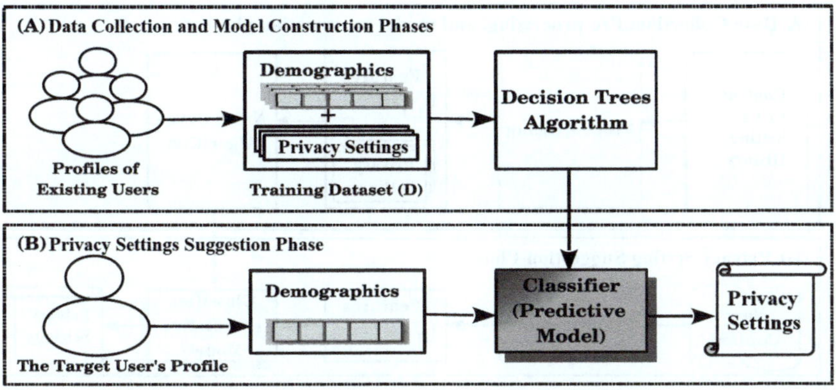

Fig. 6.2 Process of suggesting privacy policies for an individual attribute in the target user's profile [3]

training datasets, to train a series of classifiers $CL = cl_1, cl_2, \ldots, cl_{|}A|$, such that, each classifier $cl_j \in CL$ predicts the privacy settings for an attribute $a_j \in A$, $\forall j \in 1, 2, \ldots, |A|$. Finally, each time a target user registers at the SMP. We pass his/her demographics to every classifier $cl_j \in CL$, to predict (and hence suggest) the privacy settings for each attribute in the target user's profile.

6.2.3 User Content Protector (UCP)

The User Content Protector (UCP) handles enforcing privacy on the target user's generated content. The process by which the UCP suggests privacy settings for content generated by a particular target user, say $u_t \in V$ is as follows.

The first phase is the *data collection* phase, in this phase we collect the privacy settings history of the target user's content, and arrange it into a dataset Ω, where every tuple is of the form (c_i, l_j), where c_i is a content element generated by the target user u_t, and l_j is the privacy setting that u_t has specified for c_i.

The second phase is the pre-processing phase, where every content element $c_i \in \Omega$ is transformed to form a vector of features that characterise c_i. As such, Ω is transformed to form a dataset of labelled feature vectors $\hat{\Omega}$, in which every tuple is represented as (c_i, l_j), where, $c_i = \langle f_{i1}, f_{i2}, \ldots, f_{im} \rangle$ is c_i's feature vector.

The third phase is the classifier training phase where we train a Naive Bayes classifier to predict privacy policies for the target user's content. We opted to use the Naive Bayes algorithm, because it is quick to test and train especially in low processing resource settings. This is also useful and necessary in terms of predicting privacy settings for user-generated content.

Finally the suggestion phase is triggered when the target user u_t generates any new content element c_{new}. First, c_{new} is pre-processed to transform it into a feature

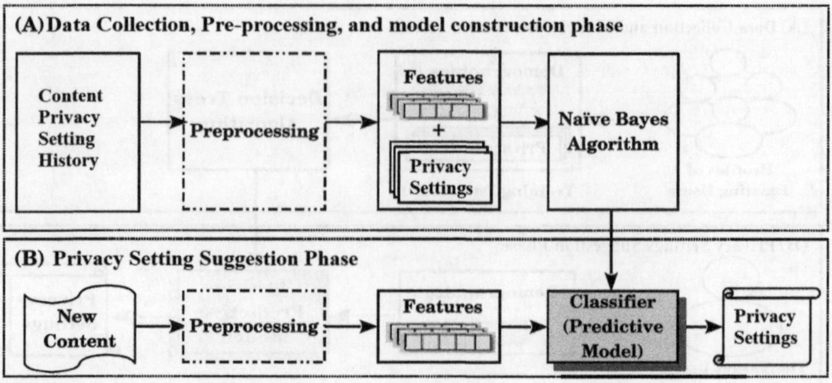

Fig. 6.3 Suggesting privacy policies for user-generated content [3]. (**a**) Data collection, pre-processing, and model construction phase. (**b**) Privacy setting suggestion phase

vector that is submitted u_t's classifier, which in-turn predicts c_{new}'s privacy setting. The predicted privacy setting is then recommended to u_t. Figure 6.3 provides an illustration of how the process of suggesting privacy settings for user-generated content works.

We now present an overview of the empirical results obtained from a proof-of-concept prototype for privacy settings recommendations [3, 4].

6.3 Prototype Implementation

In order to implement the PAP component, we used training data based on both simulated (obtained by experimentation using synthetic data) and real data obtained from another similar study [18]. The real dataset was split into ten datasets of equal size.

The simulated datasets consist of 8 profile attributes, and are structured as tables of 16 columns. Such that, the first 8 columns correspond to profile attributes, and the following 8 columns correspond to the attributes' privacy settings. The real datasets on the other hand, consist of 11 profile attributes; however, since we only have the actual values of three attributes, we structured them as tables of 14 columns, where, the first 3 columns corresponds to three available profile attributes, while the following 11 columns correspond to the privacy settings of all the 11 profile attributes.

Next, we applied the J48 algorithm (i.e. implementation of the C4.5 decision tree learning algorithm) on each of our simulated and real datasets; to train a classifier for each profile attribute in these datasets. We then performed a 10-fold cross-validation test, and recorded the accuracy of each model (that is the percentage of

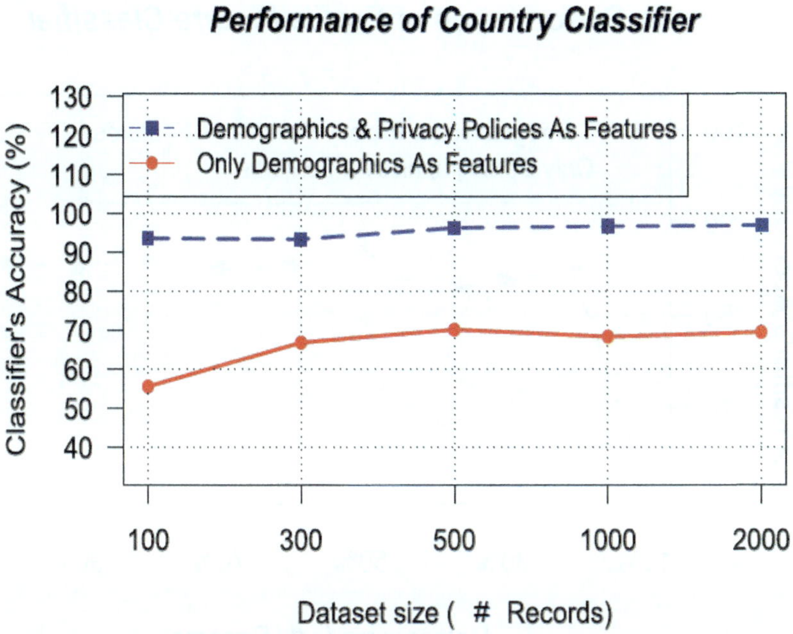

Fig. 6.4 The accuracy of the Country attribute when trained using demographic features, and extended features (Training on both real and simulated datasets using demographic and extended features.)[4]

the records for which the classifier was successful). Cross-validation results show that the classifiers trained on simulated datasets achieved an average

accuracy percentage of between 60 to 80%. While the classifiers trained on the real datasets achieved an accuracy percentage of between 60 to 70%, which is slightly lower than the classifiers trained on the simulated datasets, but not unexpected since the real dataset is less uniformly distributed than the simulated dataset.

In order to enhance the classifiers' accuracy, we extend the features used in the training phase. Instead of using only demographics as features, as we did in the initial phase, this time we used a combination of demographics and the privacy settings of the profile attributes. We then, retrained our classifiers on the new datasets, and measured their accuracy using the aforementioned methodologies. The results showed an average of 15–25% improvement in the accuracy of classifiers trained on simulated datasets, and a similar average of 17% improvement in the accuracy of classifiers trained on real datasets, as illustrated in Figs. 6.4 and 6.5 , which shows the accuracy of two selected profile attribute classifiers, plotted against the size of their training datasets.

Next, in order to implement the UCP component, we began by collecting the data necessary for building the UCP classifier. Particularly, we used Facebook's graph

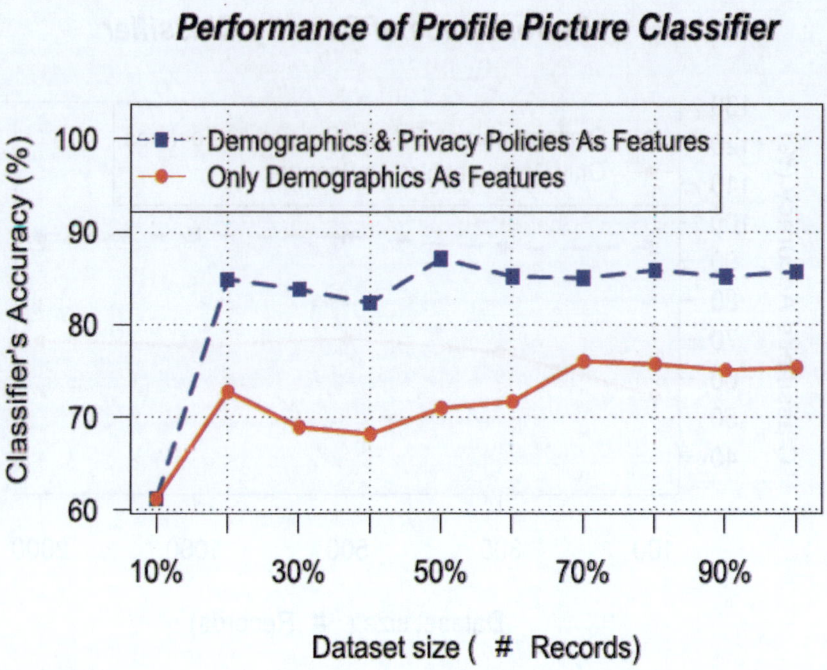

Fig. 6.5 The accuracy of the profile picture attribute when trained using demographic features, and extended features (Training on both real and simulated datasets using demographic and extended features.) [4]

Application Programming Interface (API) to download posts (i.e. user generated content) from one of the author's Facebook accounts and manually labelled each post with a suitable privacy setting. Then, in order to transform the data Ω, from a 'corpus' of text, to a dataset of labelled feature vectors $\hat{\Omega}$, we used 'text classification' pre-processing methods. Specifically, we normalised each content (i.e. post) $c_i \in \Omega$ by lower-casing letters and removing diacritical marks. Then, we stemmed the data using the Lovins Stemming Algorithm implemented by Weka[96]. Afterwards, we break-down our text corpus Ω into small units called terms or words using both word and n-gram tokenization. Lastly, we vectorised Ω such that, each content $c_i \in \Omega$ is represented as a k-dimensional vector $< c_{i1}, c_{i2}, \ldots, c_{ik} >$, where c_{ij} is a value associated with a term j in c_i and calculated each c_i, by running both the binary approach and the *tf-idf* approach to generate several pre-processed datasets $\hat{\Omega}$.

To build the UCP classifier, we divided each preprocessed dataset $\hat{\Omega}$ into a training and a validation dataset $\hat{\Omega}_{TV}$(contains 70% of the tuples or data points in $\hat{\Omega}$); and a test dataset $\hat{\Omega}_{Te}$ (contains the remaining 30%), as advised by [196]. As a next step, we employed the naive bayes, multinomial naive bayes, complement

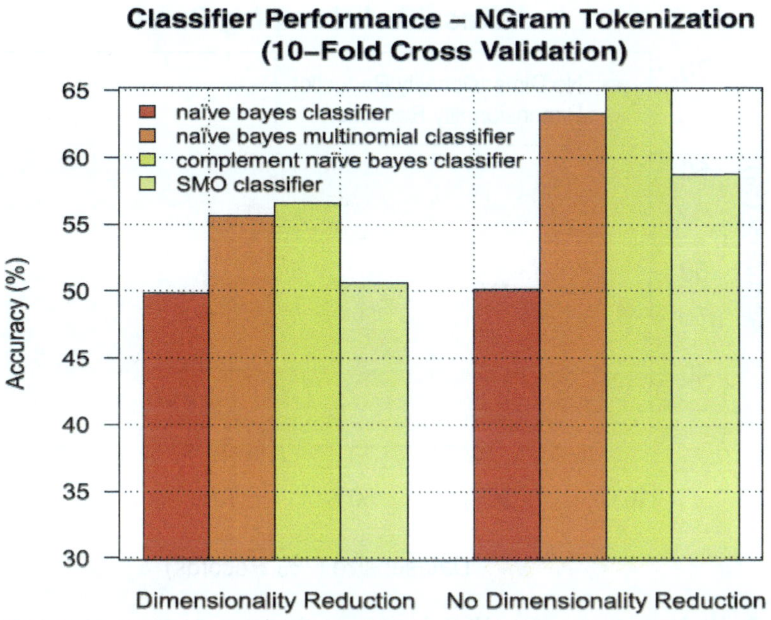

Fig. 6.6 The accuracy of the UCP content's classifier when the training data is preprocessed using 3-gram tokenisation and $tf - idf$ vectorisation [4]

Naive Bayes, and SMO (Sequential Minimisation Optimisation) or SVM (Support Vector Machines) algorithms to support the training phase using each of our training & validation datasets $\hat{\Omega}_{TV}$. Finally, we performed a 10-fold cross-validation test to measure the accuracy of each resulting classifier.

As indicated in Fig. 6.6, the results indicated that the highest accuracy was achieved when the classifier was trained using complement Naive Bayes on a dataset that was pre-processed using a combination of normalisation, stemming, 3-gram tokenisation, stop-words removal, and *tf -idf* vectorisation. This indicates that the more varied the training data the better it performs during the prediction phase, which would imply that better user recommendations are likely to arise from models trained on data belonging to users with very varied privacy settings, as opposed to a monotonic dataset that is either skewed towards very privacy-savvy users or the reverse. We then further validated this classifier on our 'untouched' testing data set $\hat{\Omega}_{Te}$confirming that the results remain stable as per the classifier's performance with an accuracy of approximately 64.2%.

We now consider some further improvements that can be implemented to improve the classifier's accuracy (Fig. 6.7).

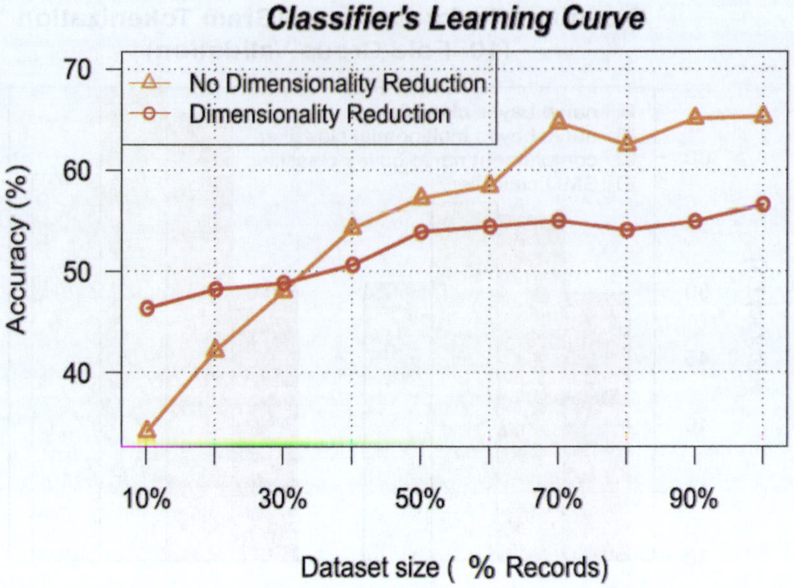

Fig. 6.7 The learning curve of the UCP classifier (i.e. the classifier accuracy plotted against the percentage of records from $hat\Omega_{TV}$ that were used in the training process [4]

6.4 Enhancing the Profile Attributes' Classifiers Accuracy

In an effort to further enhance the accuracy of the PAP classifiers, we performed extended the PAP classifiers to include, in addition to the decision tree learning algorithms such as the J48 algorithm, other algorithms such as the BFTree, DecisionStump, FT, J48, J48graft, LADTree, LMT, NBTree, Random-Forest, RandomTree, REPTree, SimpleCart, and Naive Bayes algorithms. Each classification algorithm (a total of 13 algorithms), was applied to each of our real and simulated datasets, to train a classifier for each of the profile attributes within these datasets. We then performed a 10-fold cross validation test, and recorded the accuracy of each classifier as shown in Figs. 6.8 and 6.9, where the "learning curves" of two "selected" attribute classifiers are depicted.

As shown in Figs. 6.8 and 6.9 the classification algorithms record approximately the same accuracy for the profile attributes selected with respect to the training datasets. In order to determine which algorithms out performs the others we evaluated the average accuracy of each algorithm across all training datasets, and then averaged the results across all profile attributes. Each algorithm was then plotted according to its average of the average accuracies for all the classifiers, as shown in Figs. 6.10 and 6.11.

In Fig. 6.10, the results show that for simulated datasets, the best performing algorithms are the BFTree, J48, J48graft, LMT, RandomForest, REPTree, and

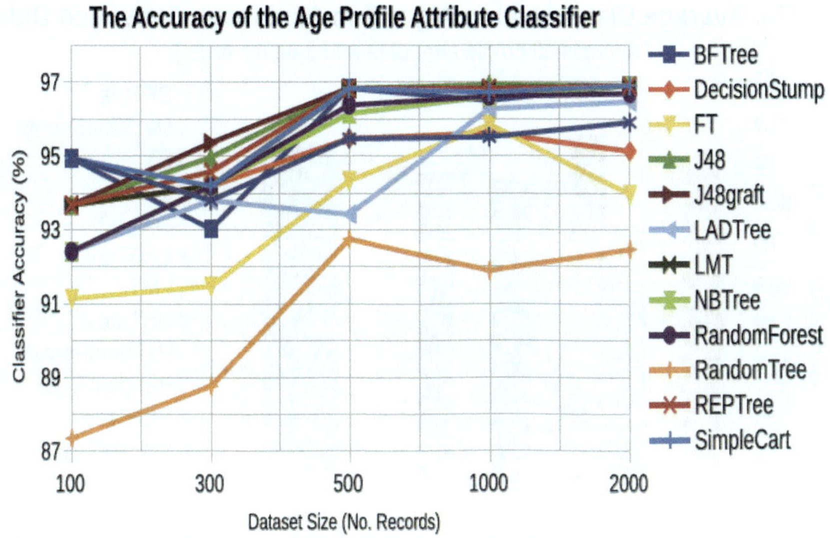

Fig. 6.8 Accuracy of the age classifier when trained on different real datasets, with various decision tree algorithms [4]

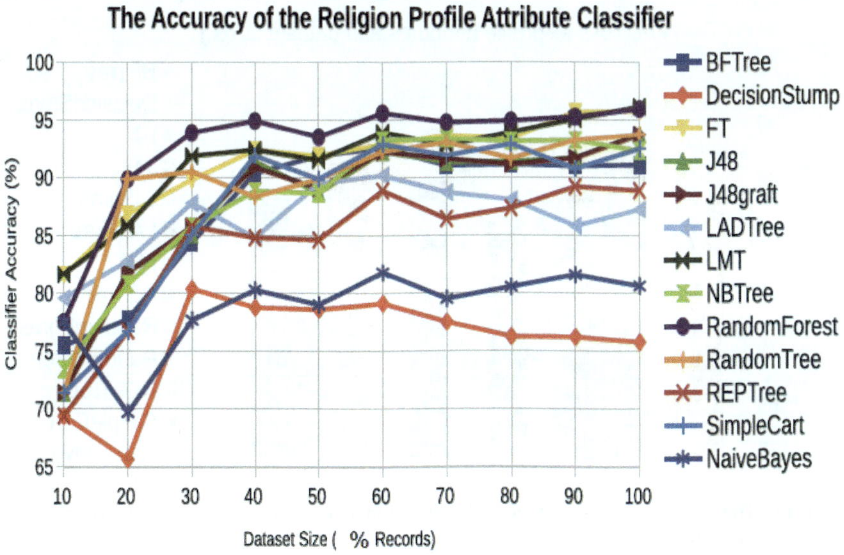

Fig. 6.9 Accuracy of the Religion classifier as its being trained on different simulated datasets using different decision tree algorithms [4]

SimpleCart algorithms, with a slight advance for the J48 algorithm. On the other hand and as illustrated in Fig. 6.11, for real datasets, the best performing algorithms are the BFTree, J48, J48graft, LMT, and Random Forest Algorithms, with a clear

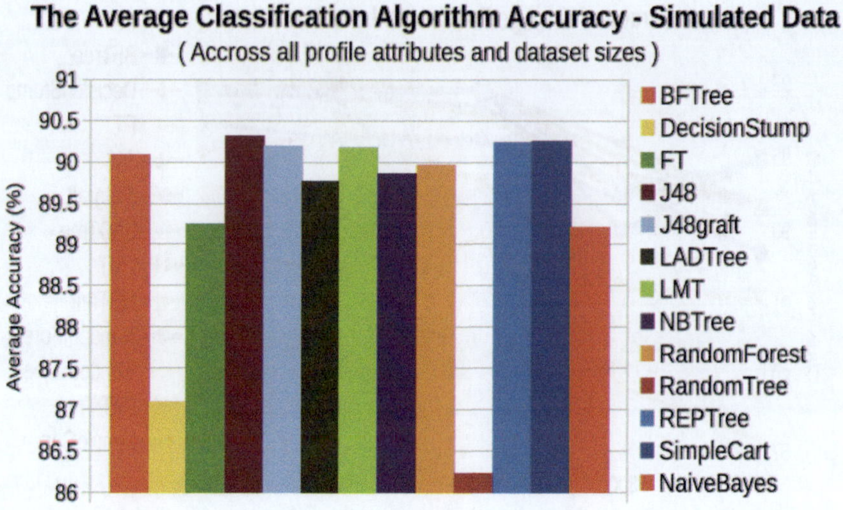

Fig. 6.10 The average accuracy of each classification algorithm across all simulated datasets and all 8 profile attributes [4]

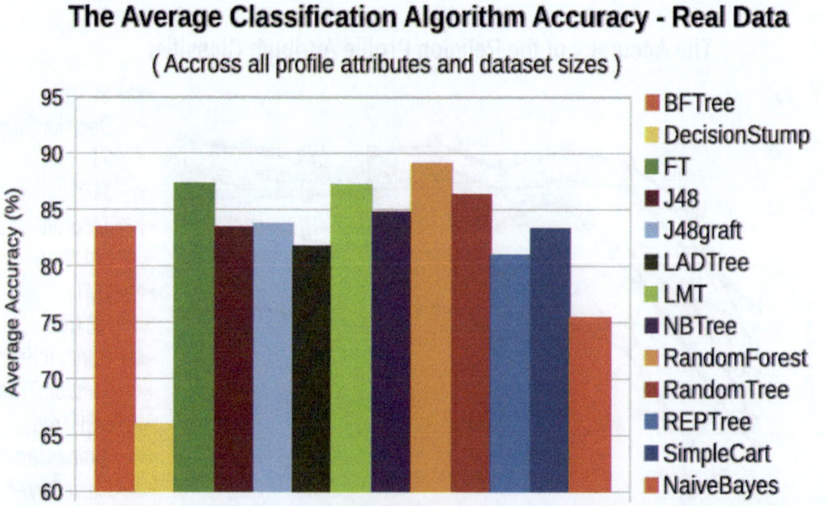

Fig. 6.11 The average accuracy of each classification algorithm across all real datasets and all 11 profile attributes [4]

dominance for the Random Forest Algorithm, with an average of 5.7% performance increase over the J48 algorithm.

Based on these results, we decided to use the Random Forest Algorithm as the main classification algorithm for training the PAP attribute classifiers. This is not only because the Random Forest Algorithm performed well on both real and simulated datasets but also because the Random Forest Algorithm is the best performing algorithm on real datasets, which is more likely to translate better to real life data sharing scenarios.

6.4.1 Attribute Privacy Settings: Considerations

Extending the feature vector used to train the profile attribute classifiers, by incorporating the privacy settings of other profile attributes can dramatically enhance the accuracy of the classifiers [3, 4]. For Instance, training the "Age" classifier on a feature vector consisting of user demographics and the privacy settings of other attributes such as "gender", and "language", results in a classifier with higher accuracy, than simply training the "Age" classifier on a features vector consisting of demographics only.

In practice however, this approach is inefficient performance-wise, because to get privacy setting recommendations for one profile attribute, say $a_i \in A$, the target users must provide the privacy settings of all the other profile attributes $A - a_i$, to a_i's classifier. This input creates a considerable overhead that is roughly equivalent to configuring privacy settings manually. As such studying the positive impact of extending feature vectors, on the accuracy of the profile attribute classifiers is an important consideration in improving the PAP component.

To this end, we impact of the correlation between the privacy settings of different profile attributes and classifier accuracy, and test this hypothesis with a Pearson's chi-squared test of independence, with a significance level (i.e. alpha) 0.01, on each pair of profile attribute privacy settings in both our simulated and real datasets. For each of the five simulated datasets, we performed a total of 28 Pearson's chi-squared tests, that is, one for every pair of profile attribute settings. For each of our ten (10) real datasets, we performed a total of 55 Pearson's chi-squared tests, which translates to one for every pair of profile attribute settings.

The results showed that while there is correlation between the majority of profile attributes' privacy settings some of these tested pairs did not comply with the conditions of chi-squared test. As such, to measure and quantify the correlation/association between the profile attributes' privacy settings we used Cramér's V measure[1] on each pair of profile attribute settings, in a similar manner to the approach we took to testing the chi-squared hypothesis (i.e. Cramér's V is based on Pearson's chi-square statistic). Cramér's V) given by the following equation:

$$V = \sqrt{\frac{\chi^2}{N * (min(k-1, r-1))}} \qquad \ldots (9.1)$$

[1] Cramér's V is a measure of association between two categorical variables [7].

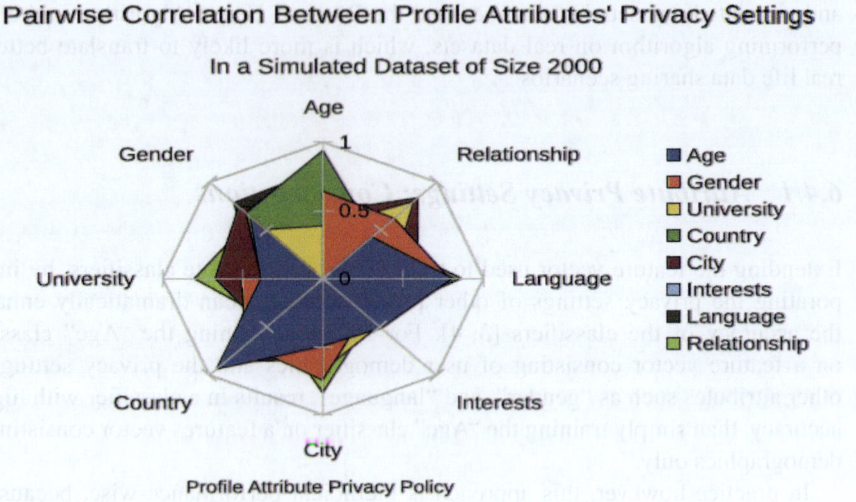

Fig. 6.12 The correlation between attributes' privacy policies in a real dataset [4]

where χ^2 is chi-squared statistic, N is the grand total of observations, k and r are the number of columns and rows respectively, such that, $V \in [0, 1]$.

We calculated Cramér's V for each pair of profile attribute privacy settings in all our real and simulated datasets. Such that, for each dataset, every pair of profile attribute privacy settings, has a numeric value in [0, 1] to quantify the correlation/association between the two.

As can be seen in Figs. 6.12 and 6.13—which shows the correlation between privacy policies of different attribute—there are many strong correlations between profile attributes' privacy policies. For instance, in Fig. 6.12 we can see a strong correlation between the privacy policies of each of the (age, country), (gender, language), and (city, relationship) profile attributes.

Similarly, Fig. 6.13 shows a strong association between the privacy policies of each of the (Location, Education), (Email, Mobile), and (Religion, Education) attributes. In addition to some medium associations between the privacy policies of each of the (Profile Pic., Gender), (DoB, Relationship), and (Name, Gender) attributes.

6.4.2 Privacy Settings Correlations

We have discussed the idea of extending feature vectors by including privacy settings of other attributes to the training dataset, and from the results can conclude that this offers one approach to achieving high accuracy yields, but that can be

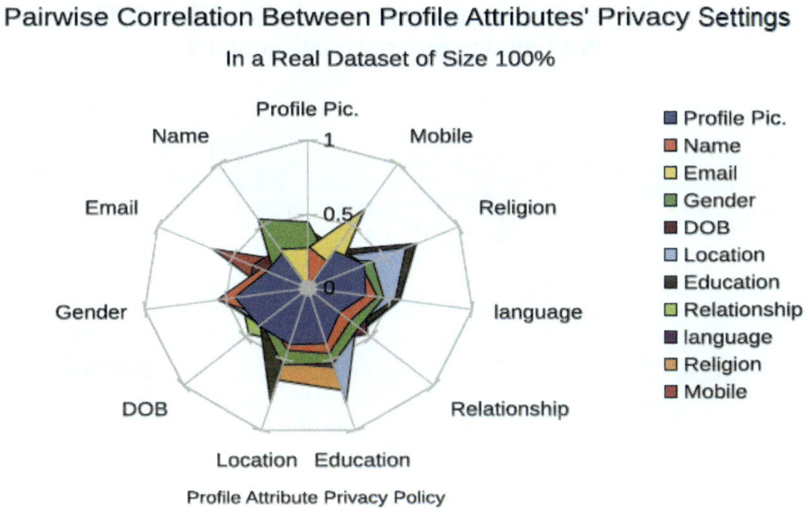

Fig. 6.13 The correlation between attributes' privacy policies in a simulated dataset [4]

inefficient in terms of placing a heavier burden on the user in terms of usability (i.e. user input associated with extending feature vectors increases).

One approach to alleviating this issue is to employ some methods to automate the task of collecting the input required to augment the feature vectors. To do so, we employed a Correlation Graph (CG), which is basically an undirected weighted graph $G = (V, E, W, \alpha)$, where V represents profile attributes, E represents correlations between profile attributes' privacy settings, W represents the strength of the correlation in E as measured by the Cramér's V measure, and finally, α is a numeric value $\in [0, 1]$. An edge $e = (a_i, a_j) \in E$ exists between attributes $a_i, a_j \in V$ if and only if the correlation between their privacy policies is greater than α. Example correlation graphs are given in Figs. 6.14 and 6.15, which show how a real correlation graph is formed from a simulated and real dataset of size 100 and 2000 respectively.

Next, we compute the Correlation Maximum Spanning Tree (CMSP) within the correlation graph to determine the spanning trees that show the strongest correlations. Figure 6.16, shows the CMSPs of the CGs shown in Figs. 6.14 and 6.15.

Unlike the CGs, the CMSTs capture the correlation/association relationships between the privacy settings of profile attributes intuitively. For instance, by observing Fig. 6.16 we can see that if we know the privacy setting of the root (i.e. gender) we can predict the privacy setting of the *city* and *language* attributes due to the correlations that exist between the two attributes. Furthermore, correctly predicting the privacy settings of the *city* and *language* attributes would in-turn enable the prediction of the privacy settings for the *country* and *relationship* attributes (Fig. 6.17).

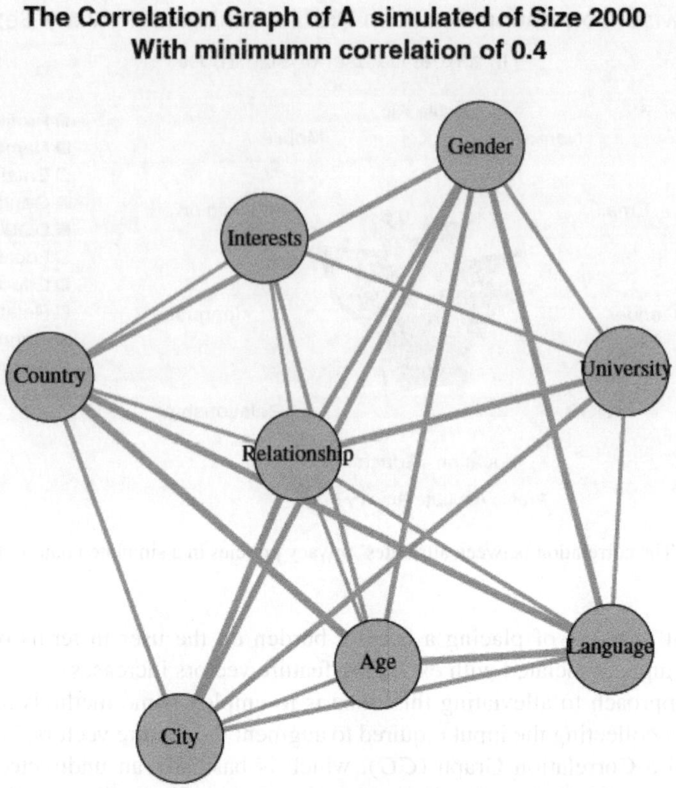

Fig. 6.14 Example correlation graph of a simulated dataset of size 2000, with $\alpha = 0.4$. [4]

This approach uses the associations within CMSTs, to add privacy settings to the feature vectors in an accumulative manner. In the classifier training phase, privacy settings are added to the root l_r or roots, when the CG has more than one CMST, of the feature vector F. The feature vector is then used to train the classifiers of the root's children. Next, the privacy settings of the root's children $l_{r1}, l_{r2}, \ldots, l_{rh}$ are added to F and these are then used to train the classifiers of root's children's children and so on until the classifiers of all nodes in the CMST are trained. During testing phase, when a new user joins the SMP, his/her demographic information is added to his/her feature vector F_t to support the automated classification process.

In order to trigger the recommendation process, the target user is asked to provide privacy settings for the root attribute l_r, and then add this (i.e. l_r) to his/her feature vector, F_t. The feature vector is then transferred to the classifiers of the root's children which in-turn predict the privacy settings for the root's children $l_{r1}, l_{r2}, \ldots, l_{rh}$ the process continues until the leaf nodes are reached and a privacy setting prediction for all attributes in the CMST has been obtained. Finally, the predicted privacy settings, for all the attributes evaluated, are recommended to the target user.

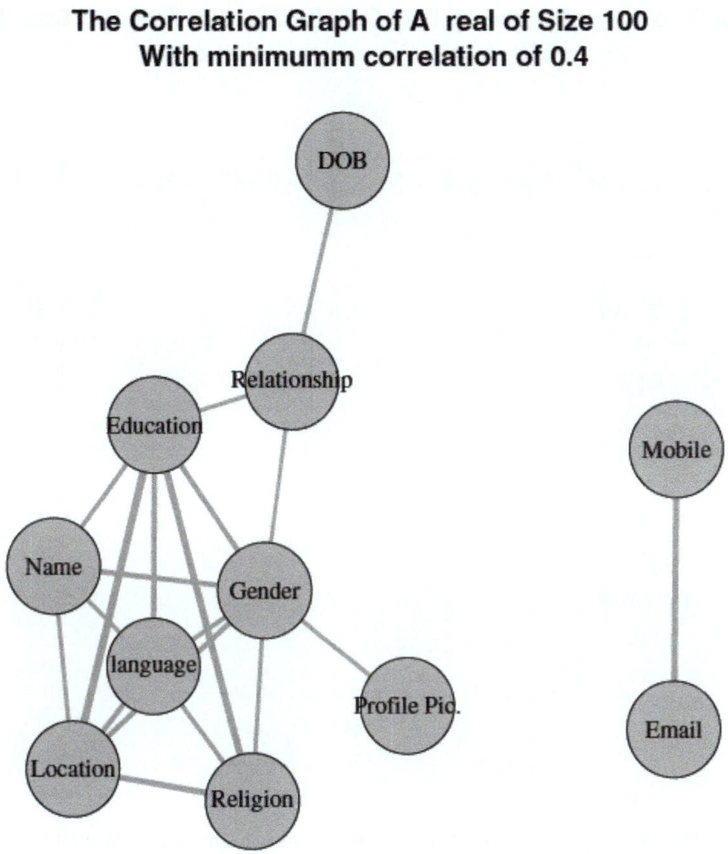

Fig. 6.15 Example correlation graph of a real dataset of size 100, with $\alpha = 0.4$ [4]

As such, target users only need to provide privacy settings of the root to trigger a recursive process that results in privacy setting recommendations for all of his/her profile attributes.

6.4.3 Experiments and Results

In order to test the validity of our proposed method, we implemented the recursive privacy settings training scheme that we described in the previous Section, and tested it on both our simulated and real datasets. We employed several open-source software packages and libraries such as, R (a statistical analysis package [75]),"GraphStream", which is a Java-based graph manipulation and visualisation library, and Weka's Java Application Programming Interface (API) [96].

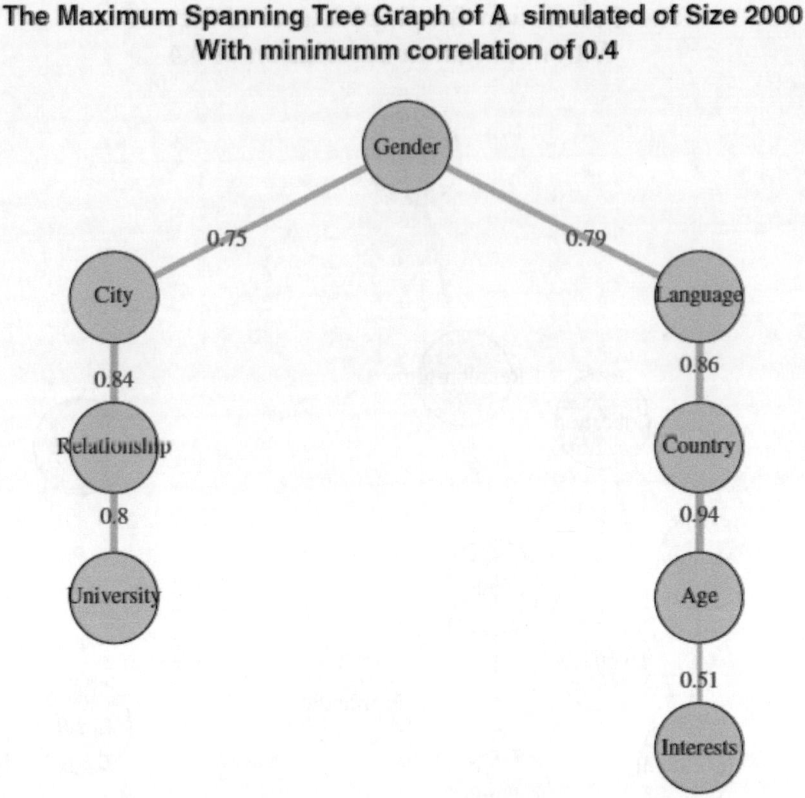

Fig. 6.16 Maximum spanning tree depicting correlations in a simulated dataset of size 2000, with $\alpha = 0.4$ [4]

We used R to build the CGs, find the CMST(s) within each CG, and store the resulting CMSTs in the standard GML format. The GraphStream library was then used to read the previously stored CMSTs, and perform operations on the CMSTs, such as, finding subtrees, and traversing the CMSTs. Lastly, we used Weka's Java API to modify the training datasets, and to train the profile attribute classifiers.

Next, after training our classifiers using the recursive, we performed a 10-fold cross-validation test to measure the accuracy the classifier in order to compare their accuracy with the accuracy of the classifiers trained using the other approaches. Figures 6.18 and 6.19, show the accuracy of the classifiers of two "selected" profile attribute (one real, one simulated), when they are trained using the recursive method, compared to their accuracy (i.e. classifiers) when they are trained using demographic features, and extended features (i.e. demographics and privacy policies).

As illustrated in Figs. 6.18 and 6.19, the accuracy of the classifiers trained using the recursive method, is very close to the accuracy of the classifiers trained using extended features for both simulated and real datasets. At the same time, we note

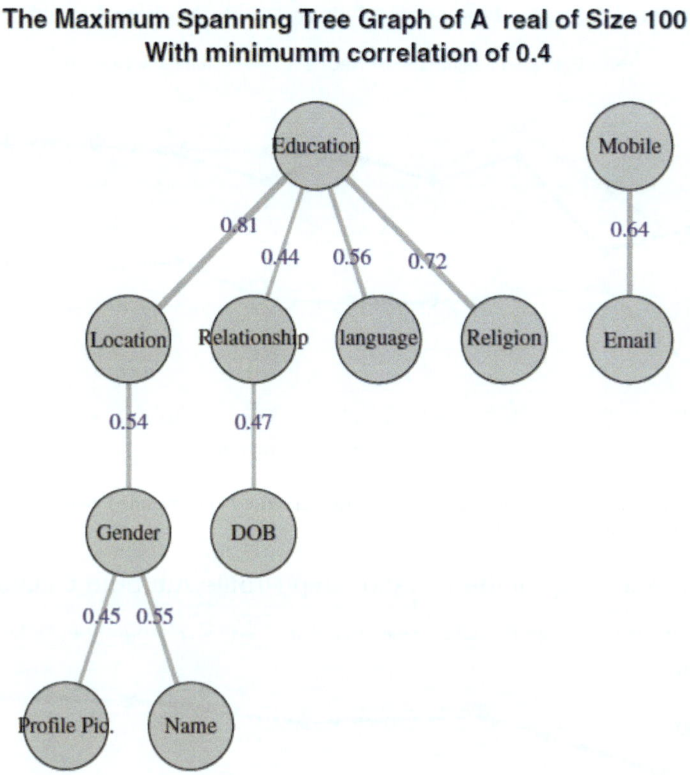

Fig. 6.17 Maximum spanning tree depicting corrections in a real dataset of size 100, with $\alpha = 0.4$ [4]

that only a fraction of the input is needed. For instance, in the simulated datasets, users need to input the privacy settings of one root attribute (i.e. gender) instead of 8 attributes, which forms an 87.5% decrease in user input. Similarly, in real datasets, users need to input the privacy policies of two root attributes (i.e. education and mobile) instead of 11 attributes, which amounts to 81.8% decrease in user input. This shows that the recursive approach to privacy settings is not only capable of achieving accuracy levels comparable to the results obtained using extended feature vectors, but it also does so much less user input.

However, it also important to note the fact that these results might be a little bit the optimistic-side, because getting a realistic idea of the performance of the recursive approach requires that the classifiers be tested on a separate testing set and not solely on cross validation. In theory, we might observe lower performance rates due to the accrued errors form previous levels in the CMST.

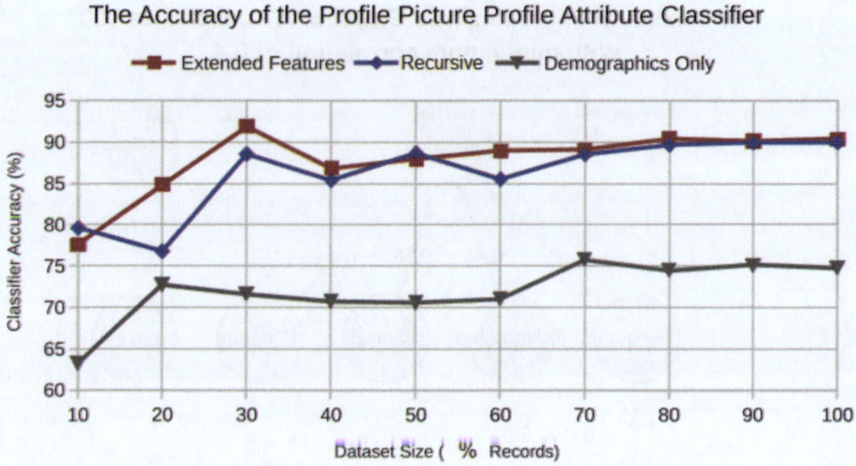

Fig. 6.18 Accuracy of the profile picture classifier based on extended features, demographic features, and the recursive method [4]

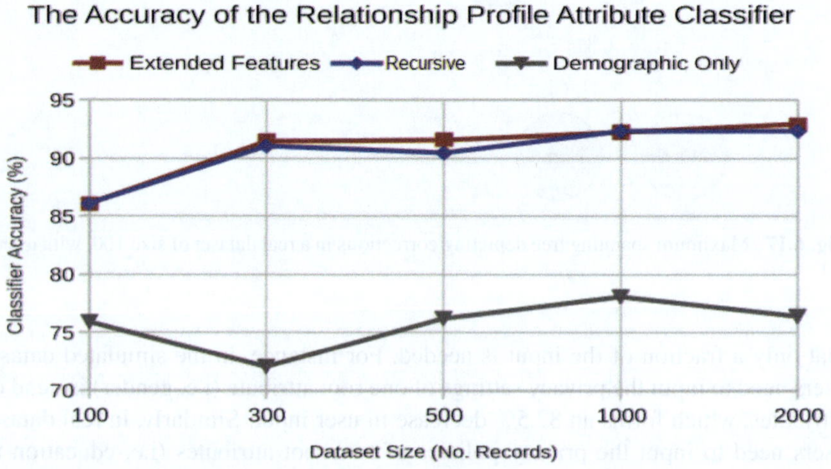

Fig. 6.19 Accuracy of the relationship classifier based extended features, demographics features, and the recursive method [4]

6.5 Discussions

The results discussed in this chapter highlight the nuances between subjective and collective privacy. While we have discussed privacy settings recommendations in relation to single users, a further dimension to this work is looking at how individual privacy settings impact on collective privacy settings. Other aspects to consider include the mental models behind user privacy decisions and how automation can alleviate such issues. However, with automation also comes considerations for how adversarial manipulations might leverage privacy aids to misled users into making unsafe privacy decisions [8, 166, 244].

In the next chapter we look at how other machine learning mechanisms such as outlier detection can be leveraged to support generating privacy preserving data, but also in enabling machine learning mechanisms to auto-detect personal information in data.

Chapter 7
Identifying Personal Information in Textual Data

In the previous Chapters we discussed anonymisation as means of transforming data to enable privacy preserving data analytics and also considered supporting issues such as user privacy preferences and usability measures to support safe privacy behaviours and by extension, the generation of privacy preserving data. As a final topic, in this Chapter, we consider some of the issues that arise in discovering personal information in unstructured data, specifically in textual data.

Unstructured data such as textual data appears on various online platforms such as social media posts, email communications, forums, and chatrooms. Data sharing on such platforms has however, raised concerns about the exposure of personal data and the potential threat this can have on the security of identity management schemes in addition to serving as an enabler for online harms. Furthermore, digital privacy legislation such as GDPR [48, 216], stipulates that organisations handling personal information must ensure that the data collected is protected at all times. As such, efficiently identifying personal information instances in data, is a cornerstone to ensuring compliance.

Discovering personal information in unstructured textual data is however, a challenging problem. Two reasons for this are that, (1) the textual is unstructured in that it is comprised of irregular grammatical structures, colloquialisms and slang; and (2) unstructured textual data consists of sparse representations of similar text elements which limits the availability of labeled data for training and by extension, the accuracy of prediction (detection) models.

In this Chapter, we discuss two approaches to enabling PII discovery in textual data, namely *Outlier Detection* and *Transformer-Based Deep Learning Models* for natural language processing to support the detection of explicit PII in unstructured textual data. With outlier detection, we study the problem of PII discovery from the viewpoint of characterising the elements of personal information as outliers or abnormal occurrences that possess the ability to provoke privacy disclosures. As a proof-of-concept, we compared the performance and accuracy of five (5) unsupervised outlier detection models namely, LOF, DBSCAN, iForest, OCSVM,

A. V. D. M. Kayem, *Enabling Privacy Preserving Data Analytics*, Advances in Information Security 92, https://doi.org/10.1007/978-3-031-93906-8_7

and SUOD. The performance comparison results indicate that iForest offers the best performance and prediction accuracy with an ROC AUC value of 0.89. To improve the prediction accuracy, we employ transformer-based deep learning models and use labelled data to support the training process. Through a comparative analysis of transformer models, such as BERT, RoBERTa, and XLNet, and an annotated a PII dataset containing 4.19 million tokens; we note that, XLNet achieves the best False Negative rate of 0.059 and a masked accuracy of 0.90. Comparative fine-tuning and evaluation on the models, indicates that BERT outperforms the other models with a 0.92 macro F1-score, a 5% improvement from baselines.

7.1 Rationale

Shared data in the form of free (unstructured) textual data appears in many scenarios and applications on the Internet. As we discussed in Chap. 6, text on social media, forums, blogs, and wikis oftentimes contains a considerable amount of personal information that is either explicitly or implicitly stated. In fact, Gandomi and Haider [83] postulate that statistics indicate that textual data constitutes approximately 95% of all unstructured data in online settings and that sharing textual data can inadvertently lead to sensitive information disclosure. In many cases, such disclosures can happen without either the users concerned or the data owners being aware of it [152, 162].

As mentioned before, a key challenge that appears in discovering personal information in unstructured textual data is that the data does not offer uniform characteristics to facilitate labeling. This is primarily due to the fact that in unstructured shared textual data, similar text elements are sparsely represented, and (especially in settings such as social media) the data generally does not follow standard grammatical structures. As such, the data very closely resembles high dimensional data which we discussed in Chap. 4. However, unlike standard high dimensional data which can be represented using a structured relational data model, unstructured data is not adaptable to the relational model. This as such limits the availability of labeled data for training and the accuracy of personal information discovery. Furthermore, when the discovered personal information elements are transformed (e.g. by masking, generalisation, and/or suppression), as happens with high dimensional data, high levels of information loss are incurred. Therefore personal information discovery in textual data must be used to transform the text based on context rather than on individual personal information elements. We now discuss, an approach to doing so, using outlier detection.

7.2 Outlier Detection

Until recently, outlier detection was studied primarily in relation to structured numerical data which lends itself well to clustering algorithms where similarity can be evaluated in terms of a scoring mechanism [12, 99, 112]. The main idea was to use outlier detection to support operations such as data cleansing to enable training and testing models based on parametric statistics. Other applications include detecting cyber-security violations such as in intrusion detection systems which are centered around detecting unusual and/or malicious activities related to operating systems' calls, network traffic, and fraud detection [11, 33, 88]. Recent works also highlight the fact that approaches such as deep feature extraction using neural networks [41] and generative neural networks [134] can also be used to predict outliers.

7.2.1 Outlier Detection in Unstructured Data

For the purpose of this work, we describe an outlier as a data point or observation, that appears to deviate significantly from other data points or observations that belong in the same sample from which the data point or observation in question originates [11, 25]. Definitions of outliers are typically centered around two main aspects: (1) Differences from the expected observations and (2) Rare cases with respect to some ground truth of normal expectations. Outliers and anomalies are terms that are often used interchangeably, depending on the context. However, for the rest of this chapter we will stick to the term *Outlier* as a manifestation of an unusual and typically dangerous (prone to privacy exposure) observation within unstructured textual data representations.

In relation to unstructured textual data, outliers can be classified as being collective in that individual instances contain elements (e.g. personal identifying data) which make them anomalous with respect to the entire dataset. For simplicity, our discussion in this Chapter is center primarily on personal identifying data (attribute values) that are both explicit and implicit in nature. We discussed in Chaps. 2–4, the ways in which explicit identifiers are represented in datasets. While potential quasi-identifiers are also an important aspect to consider in terms of the potential disclosure attacks that might arise due to contextual interpretations of the data, for reasons of scope we do not delve into this in this Chapter. We however, do study implicit personal information as a category of personal information that might not directly result in a one-to-one mapping that results in disclosure, but that when combine with other implicit personal information and/or contextual information can result in disclosure. For instance, while DATE and TIME might be implicit personal information when published individually, and also perhaps relatively harmless, when combined with other personal information say, LOCATION, or interpreted within the body of a text, might result in disclosure. It is important to note however, that quasi-identifier discovery in unstructured textual data, is an important emerging field in the area of data privacy.

7.2.2 Mechanisms for Privacy-Preserving Outlier Detection

To determine unstructured textual data elements which can be categorised as outliers, it is important to consider the correctness of labeled data and how this impacts on the performance and accuracy of outlier detection models. Outlier detection methods can generally be classified under three categories namely, (1) Supervised Models, (2) Semi-supervised Models, and (3) Unsupervised Models.

Supervised outlier detection models work well with labeled data that serves to build predictive models. However, with unstructured textual data obtaining labeled data is a challenging process in that the labeling typically needs to be done manually or by mechanisms such as regular expressions and/or active learning [136]. Supervised outlier detection also operates based on a binary classification mechanism that benefits from using imbalanced training data (i.e. more normal data instances than abnormal ones), which is difficult to implement on unstructured data mainly because the definition of "normal" is subjective. With semi-supervised outlier detection, training data is typically based on normal representations [58] or based on unlabeled data and labeled data [55].

Unsupervised outlier detection models employ unlabeled data which is well suited to the problem of personal information discovery in unstructured data. However, it should be noted that the different unsupervised outlier detection techniques each offer specific characteristics that depending on the representation of the unstructured data, result in either better or worse performance and prediction accuracy. For instance, on the one hand, unsupervised approaches such as proximity-based, density, and cluster-based methods [37, 99, 111] handle low dimensional numerical data well but are prone to over-fitting on unstructured textual data due to assumptions about data format and challenges computing distance differences [133, 207]. Since document collections are big and comprise numerous groupings, a scaling problem arises for nearest-neighbor and calculation-based techniques. While on the other hand, angle-based vector similarity is useful in estimating divergence in unstructured textual documents that can be represented as feature vectors based on word occurrences and vector cosine similarity, but is not scalable to large datasets [230].

Cluster-based approaches handle large datasets well by underlining properties such as cluster tightness and similarity but are dependent on definitions of threshold values that are difficult to quantify and as such are not practical in terms of categorising unstructured textual data based on similarity or dissimilarity in terms of the occurrence of personal information elements in the data [63, 63]. Furthermore, identifying outliers in textual data using distance and density-based approaches are processing intensive in terms of similarity calculations [12]. Dimension reduction can address this problem, but incurs high information loss when applied to identifying sensitive data [31]. Furthermore, outlier identification approaches based on subspaces could potentially be employed to address this issue by integrating pattern analysis of local data but this approach is also processing intensive and not well scalable to very large datasets [12, 13, 113].

7.2.3 Outlier Detection and Anonymisation

To improve outlier detection accuracy with respect to unstructured textual data, the meaning of words and their contextual interpretation in relation to the text's structure is crucial [13]. This impacts on data transformation decisions that are aimed at anonymisation. For instance, when anonymising unstructured textual data, sensitive information and quasi-identifiers that occur sporadically in the data, are difficult to locate and transform effectively [103]. Sweeney [208] proposed the SCRUB technique to identify and mask named entities such as an *address, name,* or a *phone number*. Neamatullah et al. [160] proposed an anonymisation technique that locates and replaces named entities with category values using regular expressions, and look-up tables. While Vico and Calegari [224] proposed an approach to detect named entities and replace them with generic category values using a variety of natural language processing technologies. Furthermore, public initiatives such as the UK Data Archive have offered tools [219] to identify and substitute numerals and words that begin with an uppercase letter with XXX. This is on the premise that most named entities in the English language are categorisable as nouns and so quite likely to appear in text with an uppercase letter at the beginning of the word. In a similar vein, Kleinberg et al. [110] proposed a mechanism to locate named entities using machine learning and anonymise data by manually reviewing documents and changing personal identifiable information. Manual data annotations as we have mentioned before, is a time-consuming and error prone operation [136].

Furthermore, as Islam et al.[103] have pointed out, masking bears similarities to pseudonymisation (see Chap. 2), in terms of the privacy it offers [104]. While approaches such as cosine-similarity can be employed to replace sensitive information in textual data with contextual similar but privacy concealing information, as we pointed out in Chap. 1, this solution does not address the issue effectively. A further consideration is that most existing work on personal information discovery in textual data, focuses either on structured data [61, 153] or semi-structured data[153] and assumes the availability of labeled data to support training personal information discovery models.

We now discuss an approach to unsupervised outlier detection in unstructured textual data, where the aim is to isolate personal information elements in textual data by characterising them as "outliers".

Table 7.1 Examples of Explicit Personal Information (EPI) in unstructured textual data

EPI attribute	Description
NAME	Person
EMAIL	Any valid email
PHONE	Any valid phone number
ID	Any valid identity number
FAC	Buildings, airports, highways, bridges, etc
ORG	Companies, agencies, institutions, etc
GPE	Countries, cities, states

Table 7.2 Examples of implicit personal information in unstructured textual data that can be used to form attribute combinations that are characterised as quasi-identifiers

Potential quasi-identifiers	Description
FAC	Buildings, airports, highways, bridges, etc
ORG	Companies, agencies, institutions, etc
GPE	Countries, cities, states
LOC	Non-GPE locations, mountain ranges, bodies of water
PRODUCT	Objects, vehicles, foods, etc
EVENT	Named hurricanes, battles, wars, sports events, etc
WORK_OF_ART	Titles of books, songs, etc
LAW	Named documents made into laws
LANGUAGE	Any named language
DATE	Absolute or relative dates or periods
TIME	Times smaller than a day
MONEY	Monetary values, including unit
QUANTITY	Measurements, as of weight or distance
ORDINAL	"first", "second", etc
CARDINAL	Numerals that do not fall under another type

7.3 Discovering and Masking Personal Information: Considerations

In line with the idea of characterising personal information as outliers, we describe an outlier as the occurrence of an element of personal information (both implicit and explicit) in unstructured textual data.

Examples of attributes that capture explicit personal information are given in Table 7.1.

As mentioned before personal information can also be expanded to include implicit personal identifier attributes such as the one given in Table 7.2

In order to detect outliers we focus on textual data containing both explicit personal information attributes such as *name, date of birth, address, etc* and implicit personal information attributes such as *gender, location, age*. We characterise the explicit and implicit personal information attributes as named entities, and use Named Entity Recognition (NER) [155] to identify and isolate them. We note that

Table 7.3 Examples of named entity categories

Feature	Description
PERSON	People, including fictional
EMAIL	Any valid email
PHONE	Any valid phone number
NORP	Nationalities or religious or political groups
FAC	Buildings, airports, highways, bridges, etc
ORG	Companies, agencies, institutions, etc
GPE	Countries, cities, states
LOC	Non-GPE locations, mountain ranges, bodies of water
PRODUCT	Objects, vehicles, foods, etc
EVENT	Named hurricanes, battles, wars, sports events, etc
WORK_OF_ART	Titles of books, songs, etc
LAW	Named documents made into laws
LANGUAGE	Any named language
DATE	Absolute or relative dates or periods
TIME	Times smaller than a day
PERCENT	Percentage, including "%"
MONEY	Monetary values, including unit
QUANTITY	Measurements, as of weight or distance
ORDINAL	"first", "second", etc
CARDINAL	Numerals that do not fall under another type

NER mechanisms are typically dependent on both domain specific information and clearly identifiable features to learn reliably from available corpura that support supervised training, and so we handle this by identifying Named Entities (NE) with the help of a pre-trained transition-based parser model that employs a Stack-LSTM,[1] that augments the LSTM model with a stack pointer [69, 103, 116].

One of the main benefits of using the stack-based LSTM model is that NER can be done by detecting a single word or a collection of words that describe an entity and can classify the words into different categories. For instance, for a given set of unstructured textual data items, say a set of reviews $R_1, R_2, \ldots, R_3$ made by different users, our goal is to locate all named entities within each occurrence of R_i and compute their frequency count by category. Each of the named entity (NE) categories is considered a feature for detecting outliers. Table 7.3 provides an example of the NE categories that served as features to represent the textual data occurrences that we experimented with (see Sect. 7.4).

In order to identify the personal information elements in the textual data, one approach is to use a natural language processing library with pre-trained pipelines

[1] Stack-LSTM: Is a stack based linear state transition model that constructs portions of the input sequentially using a stack data structure and generates representations of the stack required for prediction.

such as the NLTK and spaCy libraries [204]. Attributes such as *EMAIL* and *PHONE* which have a structured format are typically annotated using regular expressions.

The PII detection mechanism works by categorising discovered PII elements as outliers. To determine if a word in a textual document is categorisable as PII, the outlier detection mechanism will seek to determine if it is categorisable as a named-entity. For instance, when an `EMAIL`, `PHONE`, or `DATE` is discovered it is labeled as a named entity. To support the outlier detection process, we employ a framework based on a library of algorithms each of which computes outliers (PIIs) and we select the most accurate result generated. As we discuss, in the next Section, in our proof-of-concept implementation we employed five (5) outlier detection models namely: LOF, DBSCAN, iForest, OCSVM, SUOD.

Once the PII have been discovered, we can then employ a transformation mechanism to ensure privacy. One approach is to substitute the named entities with pseudo-values that are semantically similar but different. However, we note the context plays an important role in the effectiveness of such a transformation approach in ensuring privacy. On the basic level simply substituting semantically similar values raises a similar problem to that encountered in pseudonymisation approaches to ensuring data privacy.

One approach to handling masking, though, is to generates entities of a similar kind and maintain a hash-table lookup approach to produce consistent masking values that translate to the same masking value each time a particular type of named item is discovered. Semantic similarity, can then be computed based on comparing word embeddings, by evaluating the distance between the original and anonymised textual data elements rather than relying on lexicographical similarity [132].

We now discuss the implementation and experimental results.

7.4 Implementation and Experimental Results

As an example of personal information detection in textual data, using outlier detection, we consider an outlier detection mechanism that is applied to an AirBnB review dataset with comments based on properties in the area of Berlin, Germany. The dataset contained 410, 291 reviews including spam[2] and we considered comments written in English only, for a total of 253, 908 reviews. We could have used a larger dataset, but chose a smaller one due to compute processing limitations.

A sample dataset is given in Fig. 7.1.

To support our outlier detection scheme, and as a case study, we study the performance of five unsupervised outlier detection models namely, LOF, DBSCAN, iForest, OCSVM, and SUOD. These algorithms were selected based on the fact that they are state-of-the art low processing outlier detection algorithms and have been shown to be efficient in dealing with classifications involving unstructured

[2] AirBnB Dataset.

	listing_id	id	date	reviewer_id	reviewer_name	comments
0	3176	4283	2009-06-20	21475	Milan	excellent stay, i would highly recommend it. a nice flat in a very nice area. Britta provided clear instructions in securing the keys, etc. Thanks again.
1	3176	134722	2010-11-07	263467	George	Britta's apartment in Berlin is in a great area. There are numerous fantastic Restaurants and Bars to suit every taste, easy access to supermarkets and public transportation. The apartment is also surprisingly large. I would surely stay there again.
2	3176	144064	2010-11-24	76726	Patricia	Fantastic, large place in good location. Only a short tram ride to Alexanderplatz U-S Bahn, with connections across Berlin. Britta was quick to reply to all messages and gave very clear instructions regarding directions, key pick up and things to know about the place. Try Gargerin cafe around the corner for lovely and cheap continental breakfast. I definitely recommend this apartment!
3	3176	156702	2010-12-21	291657	Benedetta	L'appartamento di Britta è molto largo carino confortabile è in un quartiere molto tranquillo e comodo perchè ci sono tanti supermercati e bar dove poter mangiare. Siamo stati benissimo spero di ripetere la bellissima esperienza ... magari in estate!\r ciao Benedetta
4	3176	165048	2011-01-04	279789	Aude	We went in Berlin for the new year eve. The appartment of Britta was really amazing, very chic, cool and in a good district !\r You should go there !

Fig. 7.1 Sample AirBnB data

Table 7.4 Execution time comparison for base models [103]

Model	n_jobs=1	n_jobs=-1
LOF	3.51 $s \pm$ 274 ms	2.32 $s \pm$ 92.9 ms
DBSCAN	8.54 $s \pm$ 799 ms	3.74 $s \pm$ 346 ms
iForest	5.23 $s \pm$ 485 ms	4.2 $s \pm$ 109 ms
OCSVM	7 min 23 $s \pm$ 8.99 s	7 min 2 $s \pm$ 24.8 s
SUOD	5 min 32 $s \pm$ 1 min 19 s	4 min 46 $s \pm$ 16.6 s

textual data [33]. Further on-going work that we currently looking involves using deep outlier detection [33, 168]. For the data each of the models is evaluated in terms of performance and classification accuracy and the one with the best result selected in each case. In line with our goal of anonymisation to generate privacy preserving data, the discovered personal information is transformed using masking, generalisation and/or suppression.

Table 7.4 shows the execution time of the five unsupervised models. In the LOF and DBSCAN models we note that since the density calculation depends on the dimension of the dataset, dimension reduction helps improve performance significantly in comparison to the other three models. However, both the LOF and DBSCAN models do not scale well in terms of accuracy of personal information discovery in unstructured textual data. iForest incurs a high performance cost, but scales well to growing data sizes and, due to the isolation property, shows better performance than the density-based approaches. .

Table 7.5 illustrates the outlier score threshold, precision, recall, F1-score, Receiver Operating Characteristic (ROC) and Area-Under-the-Curve (AUC), and precision-recall AUC score for the five unsupervised models. In terms of evaluating the models to determine which is best suited to personal information discovery in a given unstructured textual representation, we use recall because we interested in reducing the false negative value (i.e. having a high accuracy rate in terms of identifying personal information correctly based on the information drawn during the training process). Table 7.5 shows that based on the recall and F1-score values, SUOD, iForest, and OCSVM perform well with recall values of 0.70, 0.69, and 0.68,

Table 7.5 Evaluation matrices for the models [103]

Model	Threshold	Precision	Recall	F1-score	ROC AUC	PR AUC
LOF	1.00	0.84	0.33	0.47	0.61	0.55
DBSCAN	–	0.90	0.18	0.30	0.58	0.51
iForest	0.00	0.73	0.69	0.71	**0.86**	**0.78**
OCSVM	500.77	0.74	0.68	0.71	0.79	0.73
SUOD	−0.16	0.75	**0.70**	**0.72**	0.84	0.77

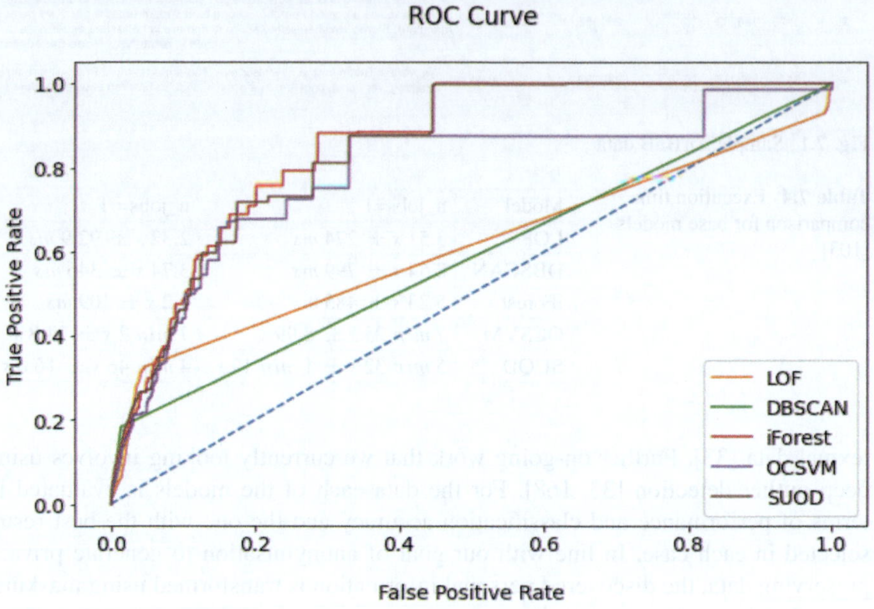

Fig. 7.2 ROC curve [103]

respectively, and F1-scores of 0.72, 0.71, and 0.71 respectively. On the other hand, LOF and DBSCAN perform worst, with recall values 0.33 and 0.18, respectively.

Figure 7.2 shows the TPR (True Positive Rate)/recall versus the FPR(False Positive Rate) at various outlier score thresholds. We note that iForest model performs best with a Receiver Operating Characteristic (ROC) and Area-Under-the-Curve (AUC) of 0.86, followed by SUOD and OCSVM. LOF and DBScan performed worst, which corroborates the result shown in Table 7.5 with respect to the recall and F1-score for the iForest model.

Figure 7.3 and Table 7.5 also show results of the precision-recall curve, indicating that iForest performs best with a precision-recall Area Under the Curve (AUC) of 0.78, followed by SUOD, OCSVM, LOF, and DBSCAN.

We trained our Word2Vec embeddings on the Common Bag of Words (CBOW) pre-trained model for performance efficiency and accuracy based on representations of words that occur frequently within a text, and isolating ones that don't being

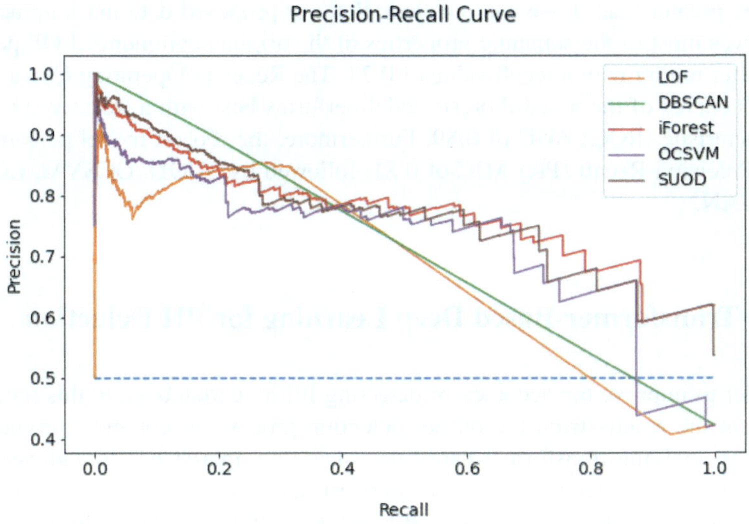

Fig. 7.3 Precision-Recall curve [103]

potentially outlierly. The resulting word embeddings are used to calculate document similarity by measuring the cosine angle. Ideally we want to have similarity values between original and anonymised documents closer to 1.

Using the named entities in Table 7.3, we pre-trained an NER system to identify named entities and calculated their frequency count by category and applied dimension reduction using Principal Component Analysis (PCA) and Singular Value Decomposition (SVD)to reduce the sparsity of the feature vectors.

We used domain knowledge to determine outlierliness in terms of rarity of occurrence. For instance, (1) If `EMAIL` or `PHONE` is present as a named entity, we consider the comment to be an outlier (personal information), and (2) likewise, for the presence of `PERSON` or `ORG` with other named entities.

$$v_i > 0;\ i \in \{EMAIL, PHONE\}$$

$$v_i > 0\, and\ v_j > 0;\ i \in \{PERSON, ORG\};\ j \notin \{PERSON, ORG\}$$

Here v_i is an element of the feature vector and i represents a category of named entities.

During the data masking step, we substitute the named entities from the outlier comments and use these named entities to generate document embeddings. This is in a bit to avoid the remaining terms from affecting the embeddings that remain unchanged in both original and transformed comments. The results show that 50% of the anonymised comments have a similarity score between 0.357 to 0.554 with a median score of 0.461 while only 7% of the transformed comments have a similarity score less than or equal to 0. As the majority of the similarity score has a value greater than 0, we can conclude that our proposed data masking approach preserves most of the semantic properties of the original comments. LOF performs best after tuning with a recall value of 0.74. The Receiver Operating Characteristic (ROC) curves of the tuned iForest model performs best with a Receiver Operating Characteristic (ROC) AUC of 0.89. Furthermore, the iForest model performs best with Precision-Recall (PR) AUC of 0.81, followed by SUOD, OCSVM, LOF, and DBSCAN.

7.5 Transformer-Based Deep Learning for PII Detection

In order to improve the accuracy of detecting PII in textual data, in this section we build on the results from the outlier detection process, to conduct a comparative study of applying transformer-based deep learning approaches to addressing this problem. We expanded the scope of our test datasets to include the Enron email (completely unstructured) and WikiPII (semi-structured) datasets, to ensure variability in terms of the data analysed. Both datasets were then combined to form a hybrid dataset (HPII). Figure 7.4 provides an overview of how the data was collected and prepared to form the three datasets we used for experimentation.

As a baseline, we use Da Silva's (2020) dataset [200] which was derived from the Enron email 2015 public dataset [47]. The dataset includes approximately 510,000 emails structured using both semi-formal and informal language, obtained from roughly 150 users. The PII in the dataset includes names,emails, and phone numbers. Figure 7.5 provides an example email.

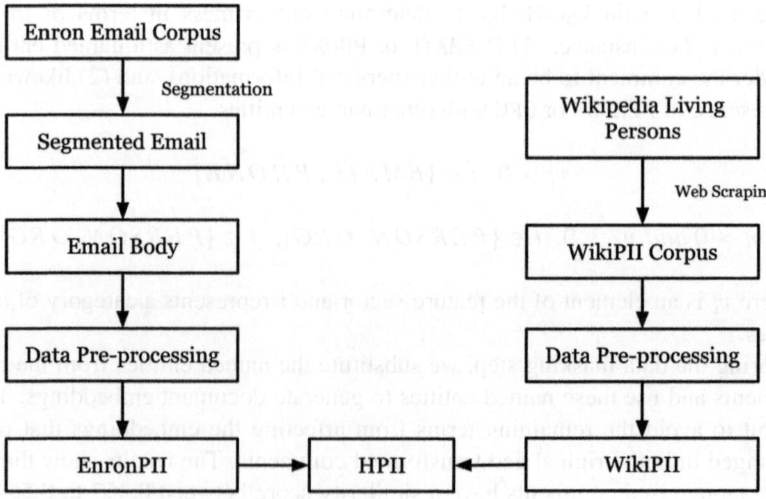

Fig. 7.4 Dataset preparation process [197, 198]

```
Message-ID: <9496454.1075855378828.JavaMail.evans@thyme>
Date: Fri, 27 Apr 2001 13:02:00 -0700 (PDT)
From: phillip.allen@enron.com
To: tim.heizenrader@enron.com
Subject:
Cc: tim.belden@enron.com
Mime-Version: 1.0
Content-Type: text/plain; charset=us-ascii
Content-Transfer-Encoding: 7bit
Bcc: tim.belden@enron.com
X-From: Phillip K Allen
X-To: Tim Heizenrader <Tim Heizenrader/PDX/ECT@ECT>
X-cc: Tim Belden <Tim Belden/HOU/ECT@ECT>
X-bcc:
X-Folder: \Phillip_Allen_Jan2002_1\Allen, Phillip K.\'Sent Mail
X-Origin: Allen-P
X-FileName: pallen (Non-Privileged).pst

Tim,

Can you authorize access to the west power site for Keith Holtz. He
is our Southern California basis trader and is under a two year
contract.

On another note, is it my imagination or did the SARR website lower
its forecast for McNary discharge during May.
It seems like the
flows have been lowered into the 130 range and there are fewer
days near 170. Also the second half of April doesn't seem to
have panned out as I expected. The outflows stayed at 100-110 at
McNary. Can you email or call with some additional insight?

Thank you,

Phillip
```

Fig. 7.5 Example of an email in Enron dataset [197, 198]

The content of the email was segmented to isolate the email body which we then focused on for PII annotation. A contextualized sentence was created from the metadata and prepended to the email body in order to retain email addresses. For example: `[X-From]` + " sent email from " + `[From]` + ". " + `[X-To]` + " received it on " + `[To]` + ". " + The Email Body

The an annotated dataset was derived by randomly selecting 10,473 unique emails from the Enron dataset [200], in which the aforementioned PII elements were manually labelled. This annotated dataset would then subsequently be used to support the training process.

In order to compose our WikiPII dataset, we extracted Wikipedia pages under the category of 'living persons'. Using Hathurusinghe (2020) [98] dataset and PII elements as a guideline, we extracted 482 documents annotated by BIRTH PLACE, BIRTH DATE, PARENTS, SPOUSES, CHILDREN, EDUCATION. We note that

Table 7.6 A summary of the datasets

Dataset	Documents	Sentences	Tokens	PERSON	LOCATION	PHONE	EMAIL
EnronPII	10,473	257,305	3,572,108	273,096	41,788	22,523	68,019
WikiPII	482	24,433	617,607	39,371	13,355	482	482
HPII	10,955	281,738	4,189,715	312,467	55,143	23,005	68,501

the WikiPII dataset does not contain explicit personal identifiers such as emails or phone numbers.

Finally, the HPII (Hybrid PII) dataset was obtained by combining the annotated data from both the EnronPII and WikiPII. This was done in order to evaluate the efficiency of the models we compared in dealing with both unstructured and semi-structured data.

In summary, Table 7.6 lists the number of annotated explicit PII's within each training dataset as well as the number of documents, sentences and tokens.

As shown in Table 7.7, the EnronPII dataset indicates an ER of 11.35, highlighting the relatively higher density of entities (non-PII tokens) compared to tokens. This implies that the dataset is imbalanced in terms of the proportion of PII entities and the TTR of 3.12, suggests a lower lexical diversity. By contrast, WikiPII shows an ER of 8.69, indicating a lower concentration of entities relative to tokens, and a higher TTR of 6.02, indicating higher lexical diversity.

The average token count per document for the EnronPII dataset is 341.08, and the high SD of 813.90 suggests that there is a significant variation in the number of tokens per document. In the WikiPII dataset, the average token count per document is notably higher at 1281.34, with an average of 25.28 tokens per sentence. Along with the larger document size, WikiPII has a higher PII density, featuring 111.39 PII entities per document and 3.06 PII entities per sentence.

The HPII dataset, is the largest of the three datasets, and presents an ER of 10.96, similar to EnronPII and a TTR of 3.09. The average token count per document for HPII is 382.45, with 16.73 tokens per sentence and an average of 41.91 PII entities per document as well as 3.95 PII entities per sentence.

We now discuss the performance of our selected transformer-based deep learning models for PII detection on the aforementioned datasets (EnronPII, WikiPII, and HPII).

7.5.1 Performance and Privacy Evaluation

To assess PII detection accuracy, we employed three transformer-based deep learning models namely: BERT, RoBERTa, and XLNET. We trained each model on a 70–30% test-train split of each of the datasets (EnronPII, WikiPII, and HPII).

Following an epoch impact analysis, we decided to use epoch=3 to minimise computational costs due to prolonged training, and also to avoid over-fitting (see

Table 7.7 Dataset statistics of tokens and entities [197, 198]

Dataset	ER	TTR	Token per doc ± SD	Token per sent ± SD	PII per doc ± SD	PII per sent ± SD
EnronPII	11.35	3.12	341.08 ± 813.90	15.80 ± 27.82	38.71 ± 84.90	4.11 ± 17.20
WikiPII	8.69	6.02	1281.34 ± 1967.27	25.28 ± 14.13	111.39 ± 165.06	3.06 ± 2.65
HPII	10.96	3.09	382.45 ± 916.74	16.73 ± 26.95	41.91 ± 41.91	3.95 ± 3.95

Doc Document, *Sent* sentence, *SD* Standard Deviation

Table 7.8 Effect of model performance using transfer learning [197, 198]

	Without pre-trained weights			With pre-trained weights		
Entity	Precision	Recall	F1-score	Precision	Recall	F1-score
EMAIL	0.97	0.97	0.97	1.00	1.00	1.00
LOCATION	0.58	0.51	0.54	0.89	0.91	0.90
PERSON	0.58	0.74	0.65	0.90	0.93	0.91
PHONE	0.98	0.99	0.99	1.00	1.00	1.00
Macro avg.	0.78	0.80	0.79	**0.95**	**0.96**	**0.95**

Appendix A). To assess the impact of transfer learning, we performed BERT model training on a 70–30% train-test split of the WikiPII dataset both with and without pre-trained weights (Table 7.8).

Without the pre-trained weights, for entities like EMAIL and PHONE, the model achieved high precision, recall, and F1-score, indicating robust performance in identifying these entities. However, for entities like LOCATION and PERSON, the performance was relatively lower, with precision, recall, and F1-score values ranging from 0.54 to 0.65. The model struggled to accurately identify entities that highly require contextual information. The macro average F1-score, which considers the overall performance across all classes was around 0.79.

With the pre-trained weights, there is a significant improvement in performance across all entity types. The precision, recall, and F1-score are consistently high, reaching 1 for EMAIL and PHONE. The macro average F1-score substantially increased to 0.95, indicating a more balanced and improved performance across all entity types. It shows remarkable improvement, especially in the identification of entities like LOCATION and PERSON, which had lower performance without pre-trained weights. This shows that pre-training BERT on contextual data plays a crucial role in enhancing its performance on NER, as demonstrated by the substantial improvement in PII detection when using pre-trained weights.

7.5.1.1 Cross-Validation and t-Testing

The threefold cross-validation involves dividing the dataset into three parts, training the model on two-thirds, and evaluating the remaining one-third. This process is repeated three times, each time using a different subset for evaluation. The metrics are the mean of the three repetitions. Table 7.9 presents the results of k-fold cross-validation (k=3) on different models for the three datasets (EnronPII, WikiPII, HPII) along with standard errors.

Running BERT and RoBERTa for cross-validation over the EnronPII dataset, results in a precision, recall and F1-score of 0.91, 0.93 and 0.92 respectively. XLNet performs marginally better with scores of 0.91, 0.94, and 0.93 respectively.

Using the WikiPII dataset, BERT achieves the highest precision, recall, and F1-score values of 0.96, 0.97 and 0.96 respectively. XLNet demonstrated the same

Table 7.9 k-fold cross validation on different models (k=3) with standard error [197, 198]

Dataset	Model	Precision	Recall	F1-score
EnronPII	BERT	0.91 ± 0.003	0.93 ± 0.006	0.92 ± 0.003
	RoBERTa	0.91 ± 0.003	0.93 ± 0.006	0.92 ± 0.003
	XLNet	0.91 ± 0.003	0.94 ± 0.003	0.93 ± 0.003
WikiPII	**BERT**	0.96 ± 0.003	0.97 ± 0.003	0.96 ± 0.003
	RoBERTa	0.95 ± 0.000	0.96 ± 0.003	0.96 ± 0.003
	XLNet	0.96 ± 0.003	0.97 ± 0.000	0.96 ± 0.000
HPII	**BERT**	0.91 ± 0.000	0.93 ± 0.007	0.92 ± 0.003
	RoBERTa	0.90 ± 0.007	0.92 ± 0.003	0.92 ± 0.003
	XLNet	0.91 ± 0.007	0.92 ± 0.003	0.91 ± 0.003

results as BERT but RoBERTa exhibited slightly less performance. On the other hand, with HPII dataset BERT achieved the best performance with precision, recall, and F1-score values of 0.91, 0.93, and 0.92. RoBERTa showed less metrics while XLNet demonstrated close performance to BERT.

BERT consistently outperformed other models across precision, recall and F1-score metrics for all datasets. This indicates its ability to capture complex patterns in the data. RoBERTa, while still performing well, displayed overall lower metrics compared to BERT and XLNet. The HPII dataset demonstrated more generalized results compared to EnronPII and WikiPII. This indicates that the models trained on this dataset generalize well to different contexts and could effectively handle diverse examples.

The threefold cross-validation results, utilizing 66.67% of the data for training, generally fell between the performance of models trained on a 50–50% split and a 70–30% split. This observation implies that using a larger portion of the data for training (**70–30% split**) might contribute to similar performance as a threefold cross-validation without the computational resources required for more extensive training.

We conducted a two-sample unequal variance **t-test** on the evaluation metrics (precision, recall, F1-score) of three models (BERT, RoBERTa, XLNet) across the EnronPII, WikiPII and HPII datasets for both holdout testing and cross-validation results. The variances of these two sets of results are different.

Considering a confidence interval of 95%, from the t-test results, we note that for the **EnronPII** dataset p-values for precision, recall and F1-score are 0.14, 0.28 and 0.23 respectively when the BERT model is used. These values suggest no significant difference in performance between hold-out test results and cross-validation results for BERT. Similarly, RoBERTa and XLNet model performances indicate no significant difference as well. The **WikiPII**: BERT and RoBERTa models' performances indicate no significant difference between the two evaluation methods. For XLNet, there is a significant difference in precision and recall (p-value of 0.03), but not in F1-score (0.10). Finally for the **HPII** dataset using the RoBERTa model results in a significant difference in the recall (p-value of 0.04), but not in

Table 7.10 Cross-dataset model evaluation [197, 198]

Training dataset	Testing dataset	Precision	Recall	F1-score
EnronPII	EnronPII	0.96	0.96	0.96
	WikiPII	0.92	0.77	0.79
	HPII	0.96	0.98	0.97
WikiPII	EnronPII	0.66	0.61	0.63
	WikiPII	0.93	0.95	0.94
	HPII	0.65	0.61	0.63
HPII	EnronPII	0.93	0.93	0.93
	WikiPII	0.96	0.80	0.83
	HPII	**0.88**	**0.89**	**0.89**

precision and F1-score. The BERT and XLNet models' performances indicate no significant difference.

We observed that a smaller dataset exhibits a significant difference in performances between holdout testing and cross-validation. Conversely, larger datasets show no significant differences between these two testing methods. These results suggest that we can use a larger dataset (EnronPII, HPII) and gain performance metrics similar to cross-validation but using holdout testing. Consequently, employing a 70–30% train-test split can achieve performance comparable to that of a threefold cross-validation in subsequent model training.

7.5.1.2 Cross-Dataset Model Evaluation

Table 7.10 presents the evaluation of XLNet models trained on one dataset and tested on different datasets. It provides performance metrics, including precision, recall, and F1-score, for each combination of training and testing datasets (70-30% train-test split).

The results indicate that, across various combinations of training and testing datasets, models trained on EnronPII perform well (F1-score 0.97) on the HPII dataset, which is probably due to the fact that the HPII dataset contains a large proportion EnronPII data. Furthermore, training on the HPII dataset results in an F1-score of 0.83, and also when tested on the WikiPII dataset (F1-score 0.83), indicating that training with the HPII dataset leads to a generalized model capable of performing well on new data. Overall, these observations emphasize that the **HPII** dataset is the best choice for achieving strong and generalized performance in our PII detection models.

Table 7.11 Key metrics from model performances [197, 198]

Model	Precision	Recall	F1-score	Accuracy
BERT	**0.92**	**0.92**	**0.92**	0.980
DistilBERT	0.91	0.92	0.91	0.979
BERTweet	0.91	0.92	0.91	0.979
RoBERTa	0.90	0.92	0.91	0.979
XLM-RoBERTa	0.90	0.91	0.91	0.979
XLNet	**0.92**	0.89	0.91	**0.982**
CamemBERT	0.91	0.89	0.90	0.978
MPNet	0.82	0.82	0.82	0.954
Electra	0.83	0.81	0.82	0.957

7.5.2 Comparative Model Training

We trained several additional models. The Table 7.11 presents the performance metrics of various models trained on the HPII dataset using a 70-30% split.

These transformer models exhibit high precision, recall, and F1-scores. This proves the overall narrative of effective PII detection can be achieved using deep learning, specifically pre-trained transformer models. MPNet and Electra, while still achieving reasonable scores, show lower performances compared to the other models. This suggests that the choice of these pre-training architectures may not be suitable for the downstream NER task compared to other models.

Interestingly, CamemBERT, a model pre-trained on French data, performs well on the English HPII dataset, showing its cross-linguistic applicability. BERT, RoBERTa and DistilBERT exhibit comparable performance metrics. The distilled version (DistilBERT) maintains a similar level of effectiveness as BERT in PII detection. BERTweet also achieves similar performance.

Precision is crucial as it measures the accuracy of positive predictions, indicating the model's ability to correctly identify relevant PIIs without producing false positives. The BERT and XLNet models demonstrated exceptional precision, achieving the highest score of 0.92 among all models. Following closely were DistilBERT, CamemBERT, and BERTweet, all with precision scores of 0.91.

Recall is critical, especially in scenarios where identifying all relevant entities is essential. BERT's high recall score of 0.92 indicates its effectiveness in capturing a significant proportion of true positive instances, underlining its capacity to comprehensively recognize relevant PIIs within the dataset. This reflected on other BERT based models which also emerged as top performers. On the other hand, XLNet showed a lower recall at 0.89.

The F1-score is particularly valuable as it provides a balance between precision and recall. BERT claimed the top spot with a score of 0.92. This showcases its ability to achieve both accurate and comprehensive predictions. It is followed closely by RoBERTa, DistilBERT, XLM-RoBERTa, and BERTweet each achieving a score of 0.91. XLNet and CamemBERT got F1-scores of 0.90.

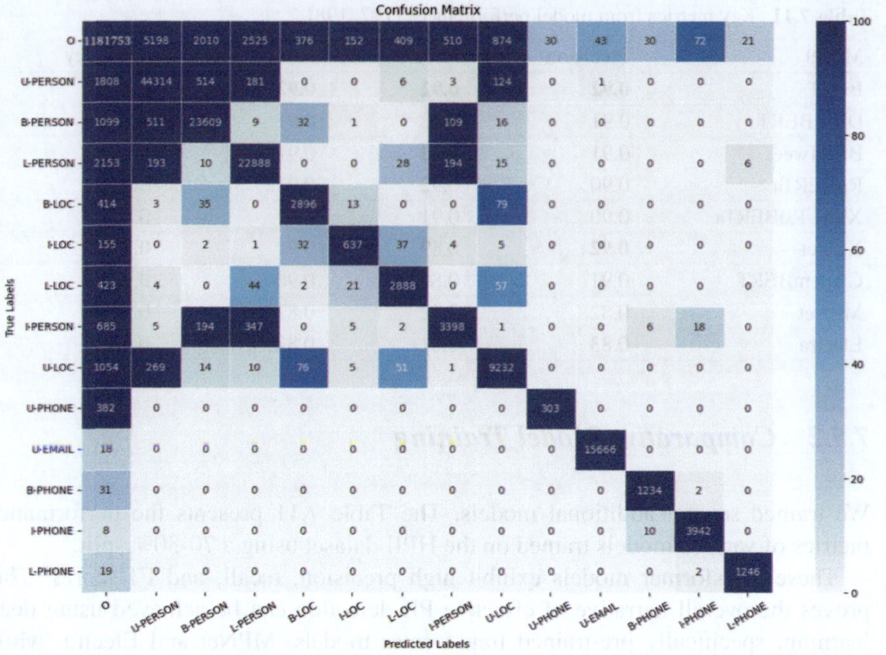

Fig. 7.6 Confusion matrix for the XLNet model [197, 198]

In addition to precision, recall and F1-score, BERT also demonstrated competitive accuracy, achieving a score of 0.980, close to XLNet's 0.982. BERT's high accuracy underscores its proficiency in making correct predictions across the entire dataset, further solidifying its overall performance.

7.5.2.1 Masked Accuracy

In the confusion matrix (Figs. 7.6 and 7.7), true labels correspond to the positive classes (PII). We can see that, the non-PII tag ("O") is most affected by misclassification. This is possibly due to difficulties the models face in detecting the correct tag with respect to the position of the tokens (beginning, inner, ending, or unit).

The masked accuracy considers only the labeled entities (PII) and ignores non-PII entities, which are marked as "O" tags. We investigated masked accuracy for the XLNet model where we achieved the highest accuracy (0.9822). The difference from the original accuracy (0.9822) is calculated as 0.0626 or 6.26%. The model still has high accuracy on the PII entities while disregarding non-PII entities.

We get masked accuracy of **0.9017** in BERT and the difference from the original accuracy (0.9800) is calculated as 0.0783 or 7.83%.

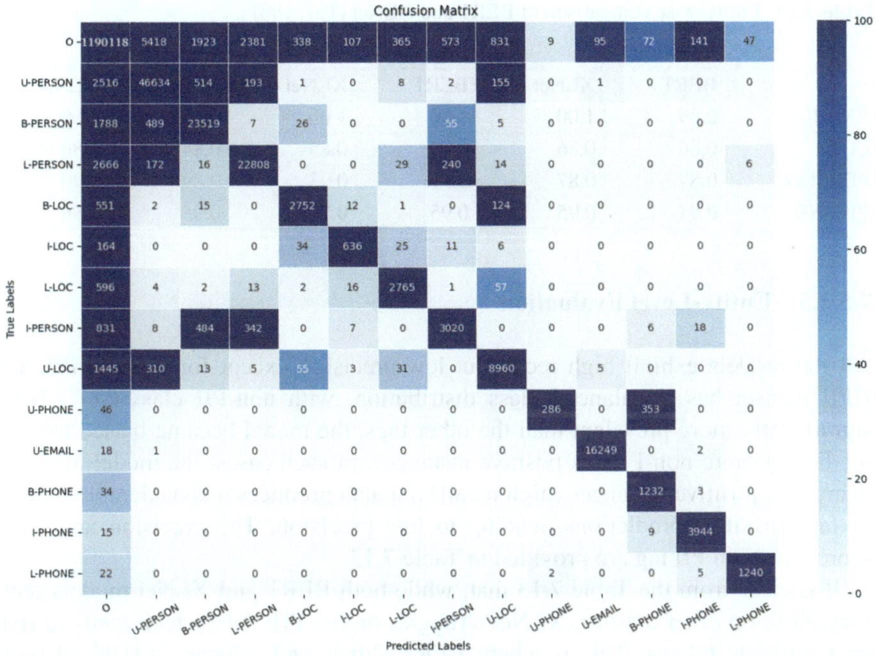

Fig. 7.7 Confusion matrix for the BERT model [197, 198]

Table 7.12 False negative rates (FNR) for different models [197, 198]

Model	FNR	Model	FNR
XLNet	0.059	XLM-RoBERTa	0.076
RoBERTa	0.067	CamemBERT	0.081
BERTweet	0.069	MPNet	0.192
BERT	0.074	Electra	0.196
DistilBERT	0.074		

7.5.2.2 Overlooked Entities

We get an insight of undetected PIIs using False Negatives (FN). FN are instances where the model incorrectly predicted entities as "O" (non-PII) when they were actually PIIs. To get a better understanding of FNs on different models, we can examine the FNR. We have calculated the FNR for the models in the Table 7.12.

XLNet has the lowest FNR (0.059) among the listed models, indicating that it is effective at minimizing the number of false negatives and correctly identifying PII entities in the text. RoBERTa and BERTweet also show reasonable performance in terms of FNR, with values higher than XLNet but still indicating a decent capture of positive instances. BERT performs relatively worse than these three and achieves a score of 0.074.

Table 7.13 Entity-wise comparison of BERT and XLNet [197, 198]

	Precision		Recall		F1-score	
	BERT	XLNet	BERT	XLNet	BERT	XLNet
EMAIL	0.99	1.00	1.00	1.00	1.00	1.00
LOC	0.86	0.86	0.82	0.85	0.84	0.86
PERSON	0.87	0.87	0.91	0.93	0.89	0.90
PHONE	0.94	0.95	0.95	**0.79**	0.95	**0.86**

7.5.2.3 Entity-Level Evaluation

All the models exhibit high recall but low precision except for the non-PII. The HPII dataset has imbalanced class distribution, with non-PII class ("O") being significantly more prevalent than the other tags, the model became biased towards predicting more non-PIIs as positive instances. In such cases, the model identifies many true positive instances (high recall) but also produces a considerable number of false positive predictions, leading to low precision. The precision and recall scores for each PII tag are provided in Table 7.13.

It is clear from the Table 7.13 that, while both BERT and XLNet models score very close on other entities, XLNet struggles on the PHONE, which tends to span over multiple tokens. This is where BERT shines and achieves a 0.95 F1-score compared to XLNet's 0.86 F1-score.

7.6 Token-Level Evaluation

The ROC-AUC assess the performance of each model in distinguishing between positive and negative classes of every token. We can closely investigate the ROC-AUC scores in Table 7.14.

We can observe that, BERT consistently performs well across all the tags, with high ROC-AUC scores, indicating good discrimination ability for identifying and classifying different parts of PIIs. All of the models perform well for identifying and classifying email entities (U-EMAIL) and single token phone numbers (U-PHONE), with relatively high ROC-AUC scores.

BERT and RoBERTa perform very well detecting the beginning and end of a name but struggle to classify the inner tokens (I-PERSON). On the other hand, XLNet can detect the beginning and inner tokens of a name but can not recognize the end (L-PERSON). It also scores low (0.44) on detecting a single token name.

BERT and RoBERTa display similar performance in contextual information as they can efficiently detect the start and end of a location entity. However, they continue to struggle with the inner tokens (I-LOCATION). XLNet also shows its strong performance in detecting the beginning but is unable to correctly classify the inner (I-LOCATION) and end (L-LOCATION) tokens. BERT and XLNet scores

Table 7.14 ROC-AUC scores for different models [197, 198]

Tag	BERT	RoBERTa	XLNet
B-PERSON	0.97	0.96	0.83
I-PERSON	**0.20**	**0.22**	0.82
L-PERSON	0.93	0.74	**0.15**
U-PERSON	0.94	0.73	0.44
U-EMAIL	0.78	0.83	0.86
B-LOCATION	0.91	0.91	0.81
I-LOCATION	**0.04**	**0.03**	**0.04**
L-LOCATION	0.50	0.53	**0.14**
U-LOCATION	0.71	0.45	0.73
B-PHONE	**0.22**	**0.11**	**0.04**
I-PHONE	0.84	**0.28**	**0.14**
L-PHONE	0.91	0.74	0.74
U-PHONE	1.00	1.00	1.00

relatively higher ROC-AUC detecting single token locations (U-LOCATION) but RoBERTa scores significantly lower (0.45) for this tag.

All models perform relatively poorly for phone number entities, especially for the beginning (B-PHONE). BERT recovers and predicts the inner tokens and the end, while RoBERTa and XLNet can only detect the ending of a phone number (L-PHONE) effectively. XLNet has the lowest scores for these tags. However, XLNet performs relatively better for the non-PII tag (O) compared to BERT and RoBERTa, as indicated by a higher ROC-AUC score.

Overall, BERT shows consistent and strong performance across various entity tags but finds difficulty tagging inside tokens. XLNet excels in predicting inner tokens but experiences challenges finding the ending. The choice of the best model as such depends on the specific requirements and priorities for different entity types in the PII detection task.

7.6.1 Computational Intensity of Advanced Libraries

The integration of the ACE library posed formidable challenges during the installation process. We encountered issues such as incompatible Spacy versions, dependency conflicts, missing optimized TensorRT on Colab, and various other module import errors. Despite our persistent attempts to resolve these issues, the ACE library proved to be unstable with Python 3.6 and later, and its compatibility with other essential packages requiring a version higher than Python 3.6 further hindered successful integration. Consequently, due to these technical impediments, we made the strategic decision to discontinue our pursuit of leveraging the ACE library for our PII detection experiments.

Our attempt to replicate the concept of embedding concatenation using the Flair library demonstrated a more viable approach. By converting the data into CoNLL

Table 7.15 Embedding concatenation with Flair on various embeddings and dataset sizes [197, 198]

Embeddings	Dataset (Size)	Training status
BERT + XLNet	HPII (70–30%)	CUDA out of memory immediately on training.
BERT + RoBERTa	WikiPII (70–30%)	CUDA out of memory in the first epoch during training.
GloVe + Flair (news-forward)	HPII (70–30%)	CUDA out of memory in third epoch during training.
GloVe + Flair (news-forward)	HPII (1–99%)	Training completes.

format, conducting pre-processing, and employing Flair's Sequence Tagger, we successfully trained a model for PII detection. However, challenges arose when working with larger datasets, particularly with embeddings like BERT and XLNet, leading to CUDA memory limitations during training.

We explored a simpler combination of embeddings, specifically GloVe + Flair, on the HPII dataset and RoBERTa + BERT embeddings on the smaller WikiPII dataset. Surprisingly, this approach also encountered the same CUDA out-of-memory error, prompting further investigation into embedding combinations. Notably, when using GloVe + Flair embeddings on a significantly reduced dataset (1–99% of HPII), the training process was completed successfully, although the precision and recall are very low due to the small amount of data (Table 7.15).

Our experiments with ACE and Flair for PII detection highlight the significant computational demands associated with building advanced models. The comparison between these two approaches underscores the intricate nature of language models and the complexities involved in enhancing the effectiveness of PII detection systems. This computational intensity acts as a safeguard for data privacy, reducing the likelihood of advanced PII detection systems being easily accessible to unauthorized entities.

7.6.2 Hyperparameter Tuning Results

We searched through the following predefined set of hyperparameters using Grid Search to optimize model performances on BERT with the HPII dataset (70–30% split).

- Learning rate: 1e-04, 1e-05, 2e-05
- Weight decay: 0, 0.01, 0.02, 0.03

The model seems already improved with learning rate 1e-04 and weight decay 0. The model achieved comparable performance on other hyperparameter combinations. With a learning rate of 2e-05 and weight decay of 0, recall increased to 0.93 at the cost of precision reducing to 0.90.

Table 7.16 Hyperparameter tuning on BERT [197, 198]

Weight decay	Learning rate	Precision	Recall	F1-score	Accuracy
0.00	1e-04	**0.92**	**0.92**	**0.92**	0.9800
0.01	1e-04	0.91	0.91	0.91	0.9796
0.02	1e-04	0.91	0.92	0.92	0.9798
0.03	1e-04	0.91	0.92	0.92	0.9799
0.00	1e-05	0.89	0.92	0.91	0.9781
0.01	1e-05	0.89	0.92	0.91	0.9785
0.02	1e-05	0.89	0.92	0.91	0.9792
0.03	1e-05	0.89	0.92	0.91	0.9778
0.00	2e-05	**0.90**	**0.93**	**0.91**	0.9791
0.01	2e-05	0.90	0.92	0.91	0.9793
0.02	2e-05	0.90	0.92	0.91	0.9790
0.03	2e-05	0.90	0.92	0.91	0.9793

It can be observed that accuracy increases with weight decay when the learning rate is constant. Precision, F1-score, and accuracy increase with the learning rate when weight decay is constant.

Table 7.16 provides insights into hyperparameter choices. Further fine-tuning experiments can be conducted to narrow down the optimal hyperparameter values.

7.7 Key Insights

The analysis provides important insights into the performance of PII detection models across various datasets and experimental conditions. Key observations include the existence of imbalanced PII distributions in our named entity datasets and the strategic selection of the third epoch for model training to balance metrics and prevent overfitting. Our experiments reaffirm that transfer learning plays a critical role in transformer models and significantly enhances performance. The 70–30% train-test split consistently produces optimal metrics and t-test results indicate that larger datasets, such as EnronPII and HPII, can achieve comparable performance metrics to cross-validation without the associated computational demands.

The models demonstrate strong performance with the mixed content HPII dataset, making it a compelling choice to achieve robust and generalized results. However, the analysis also reveals challenges, such as misclassification of the non-PII tag and difficulties in predicting entity positions, especially the inner tokens. Experiments with ACE and Flair library underscore the significant computational demands associated with these advanced models.

Notably, BERT demonstrates competitive 0.92 macro average F1-score, highlighting its proficiency in detecting correct PIIs. These findings highlight the strategic considerations involved in optimizing PII detection models to achieve robust and generalized performance.

7.8 Discussion

In this Chapter, we discussed two approaches to discovering personal identifying information (both explicit and implicit) in unstructured textual data, by characterising personal information as outliers and by employing transformer-based Deep Learning to detect explicit PII in textual data. By using outlier detection supported by named entity recognition it is possible to detect outliers (personal information) in unstructured textual data, in a performance efficient manner by using standard unsupervised outlier detection models. The results from the proof-of-concept implementation and experimentation show that the isolation Forest model predicts outliers with a Receiver Operating Characteristic (ROC) and Area Under the Curve (AUC) score of 0.86 and a recall value of 0.69. Detected outliers can be masked to anonymise but preserve semantic similarity to enable privacy preserving data analysis. Our similarity scores comparing the original and anonymised text show a median score of 0.461.

While masking can offer a solution to the problem of generating anonymised unstructured textual data, it should be noted that masking raises a similar issue to pseudonymisation, in terms of generating privacy preserving data. As such more research is needed to find more efficient and effective anonymisation algorithms for unstructured data, that preserve the semantic similarity of the data while ensuring privacy. Furthermore, as multimedia data increasingly becomes more commonplace on the Internet, it makes sense to also expand our scope in future work to study include other types of multimedia data such as image, voice, and video [135].

Other aspects worth considering in this sphere include inferences drawn from looking at combined quasi-identifiers and how these aspects might impact on both on assertions of privacy but also on the truthfulness of results drawn from the privacy preserving data [174, 175], performance [176], privacy of other types of data [135].

Regarding the use of transformer-based deep learning models we note that this has important implications for data privacy and data anonymisation, in terms automatically detect PIIs in textual data. Detecting PII in textual data can help prevent users from unknowingly sharing sensitive information and raise awareness of sharing personal information online.

The recognition of imbalanced PII distributions within named entity datasets highlights the need for more effective schemes for handling imbalanced data during the preprocessing phase. Selecting an epoch to prevent overfitting and leverage transfer learning techniques to capitalise on pre-existing knowledge is also potentially useful in handling contextual interpretations of PII occurrences in the data.

Using a 70–30% train-test split offers a practical guideline for dataset partitioning to obtain optimal results on any given dataset. The discussion on advanced libraries also highlights the importance of pre-assessing the practical implications of performance versus time-to-obtain valid results, model complexity, and computational efficiency.

BERT's macro F1-score of 0.92 underscores its proficiency in accurately detecting PIIs and indicates that BERT and related transformer models are strong candidates for PII detection tasks. Future work in could include further exploration of their capabilities in diverse contexts and leveraging work in the field of large language models.

Part V
Conclusions and Future Work

Chapter 8
Conclusions

This Chapter presents a summary of the contributions of this monograph, as well as a discussion of some possible avenues for future work. The main theme of this monograph was centred on the issue of generating privacy preserving data in an efficient and reliable manner. We pointed out, in Chap. 1, that while individual privacy decisions are subjective, to provide firms guarantees of privacy in the digital sphere, solutions must adopt a collective view. That is individual privacy decisions must be viewed in terms of their impact on the privacy of other users. As such, generating privacy preserving datasets to enable data analytics must account for a variety of considerations centered both on the privacy decisions of data owners (users) and on contextual considerations such as data formats, processing power, and data usability. Some examples of applications that can benefit from privacy preserving data include: privacy preserving machine learning tools, marketing analysis, recommendation models, healthcare applications, and digital learning platforms.

We pointed out that in order to enable privacy preserving data analytics (machine learning) not only are efficient data transformation algorithms needed but we also need these algorithms to minimise the information loss incurred due to the transformation of the data. Minimising information loss is important, because the applications that depend on such datasets need usable data to provide firm guarantees of reliability in terms of query response accuracy. In other words, the anonymised (privacy preserving) version of the data is only useful if the same (or at least a similar) query result can be returned irrespective of whether or not the data has been transformed.

To this end, we focused on the following aspects in a bid to enable privacy preserving data analytics: (1) Data representations (formats) as an important aspect in terms of deciding on what the most appropriate anonymisation mechanisms to employ in transforming data for privacy; (2) Use-case scenarios centered around constraints such as processing power, dataset sizes, as well as performance and scalability of the privacy preserving generation mechanism; (3) Mechanisms for

A. V. D. M. Kayem, *Enabling Privacy Preserving Data Analytics*, Advances in Information Security 92, https://doi.org/10.1007/978-3-031-93906-8_8

minimising information loss to maintain the veracity of the anonymised data while protecting privacy; (4) Accounting for data owner (user) privacy preferences and studying the impact of generating privacy preserving datasets; (6) Mechanisms to nudge data owners (users) to make safer and more privacy-aware decisions when sharing personal data, which is useful supporting privacy preserving data compositions; and (7) Discussing the implication of techniques such as outlier detection, deep learning, and deep outlier detection on enabling personal identifying information discovery in data.

8.1 Summary

Based on the afore mentioned considerations, in this monograph, we presented and discussed some approaches, to enable privacy preserving data analytics. We began in Chap. 2, with a review of the literature on data anonymisation mechanisms focusing particularly on the weaknesses in these schemes that enable de-anonymisations. We highlighted the challenges that arise in providing strong guarantees of privacy, and explained why sub-optimal privacy is promoted but not enough in data intensive applications on the Internet. We also considered how user intensive applications such as browsers and social media platforms raise further privacy considerations in terms of user behaviours and the resulting collection of personal data. We followed this with a further discussion of the de-anonymisation attacks that appear in social media settings and how these enable disclosures of personal information. Finally, since the Internet and digital sphere in general, raises scenarios for data compositions it was also important to discuss the implications of data representations (types) and context in the effectiveness of anonymisation methods. To this end, we focused specifically on data that contains missing or NULL values discussed some of the considerations, from the adversarial perspective, that need to be taken into account in generating privacy preserving versions of such datasets.

We built on the discussion in Chap. 2, to proceed in Chap. 3, with a case study of anonymising lossy data collected and processed in a resource constrained setting (low powered and low processing). Lossy data basically refers to data representations in which some elements of the data are missing or lost, as its name implies, due to either to the processing or data collection methods applied to the data. In resource constrained settings, such datasets typically are a result of unstable network communications and can result in very sparsely populated datasets that are challenging to anonymise effectively. In this case, the key idea is to highlight the fact that context plays an important role in the choice of approach adopted to generate privacy preserving data. We recall that the success of an anonymisation scheme depends largely on the ability to create strong similarities between the data points, so as to make potential re-identifications hard. A further issue that lossy datasets present is that lossy datasets, in resource constrained settings, are usually smaller than similar datasets composed on standard wired or wireless networks, thereby increasing the risk of re-identification. We showed that one approach to addressing this issue is to employ multi-objective anonymisation is a good approach to anonymising such datasets because the trade-off between information loss and privacy can be adjusted to match the use-case scenario. For instance, the multi-objective anonymisation approach that we proposed in Chap. 4 offers a method of anonymising datasets on low processing resource platforms without incurring high information loss. A further benefit of multi-objective anonymisation is that by weighting sensitive attribute values the risks of de-anonymisation can be alleviated.

However, on standard high speed Internet platforms, we are seeing an increased presence of large high-dimensional datasets. Examples of high dimensional datasets emerge in social media applications, fitness tracking applications, genome data, and digital education platforms. High-dimensional datasets present a challenge in that the datasets are characterised by a sparse distribution of attribute values in comparison to the number of attributes. Applying privacy preserving mechanisms such as multi-objective anonymisation to such datasets results in a high level of information loss, which is undesirable. We address this issue in Chap. 4, with a scheme designed to generate feature sets of data by breaking the functional dependencies that enable quasi-identifier (composite keys). The feature sets are in fact, subsets of the data that can be published together safely without a privacy disclosure risk. The benefit of this approach is that information loss is minimised significantly without impacting negatively on privacy. A further benefit for privacy preserving analytics is that this mechanism can also serve the dual purpose of (1.) Verifying that both current and future dataset compositions (that is modified and/or augmented versions of the dataset) are privacy compliant and (2.) Checking new datasets to quantify privacy disclosure risks.

In Chap. 2, we mentioned that skewed data is one cause of de-anonymisation and that a primary cause of this are the inferences drawn from the sensitive attribute distributions in the equivalence classes. We build on our discussion in Chap. 4, to discuss (in Chap. 5), the implications of user privacy preferences on generating anonymised datasets that are not vulnerable to disclosures due to skewed data. We consider how privacy constraints can be relaxed based on user privacy preferences and highlight the fact that while privacy might be a subjective notion in the physical world, applying this to the digital sphere requires that we give some consideration into how user privacy preferences can be translated into privacy policies that are expressive of collective privacy goals in the digital arena.

We build on this, in Chap. 6 to present a mechanism to nudge users into making safe privacy choices online. As a case study, we look at how this might be designed for a social media environment. We focus on textual and categorical data for the purpose of this Chapter, but this can be extended to deal with other types of data. The mechanism we propose in Chap. 6, illustrates how one might use automated methods to support users by suggesting (recommending) safe privacy choices to users on the Internet in general, and on social media in particular.

Chapter 7 discusses the growing prevalence of personal identifying information in the digital sphere and the implications for complementary fields such as data security. We discuss outlier detection, deep learning, and deep outlier detection as possible approaches to enabling automated personal information discovery in data, and the challenges that appear in doing so.

8.2 Future Work

As future work there are a number of aspects, drawn both from the work we have presented in this monograph but also from ongoing work in the wider field of data privacy, that should be mentioned. We discuss some of them below.

8.2.1 Nudging for Privacy

Some potential avenues for future work, would be to consider the implications of privacy from the user or human perspective. This requires studying both suggestive and pre-emptive privacy supporting mechanisms. Suggestive privacy mechanisms such as the one we presented in Chap. 6 leave the choice of action to the user. That is, privacy remains the choice and decision of each individual user. This goes in line with policies such as the democratisation of the Internet [40], where each participating party should be allowed freedom of expression and/or choice. However, suggestive privacy mechanisms also raises a number of issues in terms of data protection. For instance, several studies have shown that it is possible to subvert suggestive privacy mechanisms and that users who are heavily reliant on them can in fact be deliberately induced into error [59, 135]. Furthermore, suggestive privacy mechanisms are also reliant on personal data which raises the question of how such datasets are protected (and anonymised), and what the risks of privacy violations are based on the use of such models. This also correlates well to the discussions in Chaps. 3, 4, and 5, regarding anonymisation mechanisms and how these mechanisms can be used to support optimal privacy.

8.2.2 Supporting Privacy Decisions

In terms of preemptive privacy mechanisms, one possibility would be to build on the work we presented in Chapter to include a user content protector (UCP) component to the nudging (recommender) system and to design mechanisms to improve the accuracy of the UCP content classifiers. While Chap. 6 focused primarily on textual data, a potential avenue for future work is to study how the generalisability of the model to other content-types such as images, video, and voice content, for instance; and to incorporate more recent machine learning techniques to increase the effectiveness of the model. As mentioned in Chap. 5, some work we have conducted on pre-emptive privacy mechanisms in relation to image privacy on social media could be repurposed to support this line of work. Furthermore, social media scenarios offer a useful application scenario to conduct experiments centered on the implications of collective and individual privacy decisions with respect to jointly owned content. Sharing co-own content can raise privacy conflicts, that must be resolved by some form of a consensus [92]. Delays due to the inaccessibility or absence of user (data owner) decisions can negatively impact performance and by extension perceptions of the privacy preserving scheme [135, 152].

8.2.3 Data Transformation

We have discussed transformation measures mainly from the perspective of textual and categorical data. However, an important area that needs to be considered is that of transforming multimedia content for privacy. For instance, transforming an image by deleting or replacing a portion of it requires that we think about the implications of complex "quasi-identifiers" such as colours, background information, and even associated data on the potential inferences that might result in the disclosure of the concealed aspect. Furthermore, we are faced not only with the contention between privacy and data analytics, but also the fact that technologies such as DeepFake [147] raise new questions centered on the question of distinguishing between privacy and impersonation as well as how to achieve privacy effectively (and reliably) in a complex digital world. Other aspects such as the reversibility of masking procedures need also be accounted for in the design of effective privacy preserving solutions for such contexts.

8.2.4 Analysing Data for Disclosure Risks

Finally, it becomes important to also study outliers in anonymised datasets not only in terms of their potential for disclosure but also in terms of how they might be exploited to induce users into sharing more personal information, and/or inducing decision making tools into making erroneous decisions. This goes in line with dealing with issues such as base-rate-neglect, and the impact of adversarial manipulations to data to nudge users towards unsafe privacy choices, or false data analytics conclusions.

Since attribute compartmentation is a computationally intensive process, we present two (2) approaches to optimising attribute compartmentation namely, with Bayesian networks (Chap. 5) and parallelism via high performance mechanisms such as GPU processing (Chap. 6).

As a further solution, quasi-identifier discovery can be used to analyse already anonymised datasets to determine whether or not the datasets contain re-identification risks. This has the benefit of providing the data analytics mechanisms with a means of pre-processing datasets to ensure that they are privacy preserving before conducting analysis on the data. Since skewed data is one cause of de-anonymisations and also contributes to biased conclusions from the anonymised data, we also present an approach based on quasi-identifier discovery and transformation to alleviate this issue. An added benefit of this solution is that it allows that data analytics mechanisms (or human experts looking at the data) to guard against base rate neglect (Chap. 7) [173, 174].

Appendix A
Epoch Testing

A.1 Epoch Impact Analysis

To conduct an epoch analysis we trained, as a case-study, the BERT model on a 70–30% train-test split of the WikiPII dataset, and obtained the following results for epochs 1 to 10 in Table A.1.

The precision, recall, and F1-Score values consistently improve or remain stable over epochs 1–10. Epochs 2–5 show high performance across all metrics, which indicates stability and optimal model learning. This is highlighted in Table A.1 (see Epoch 3).

Table A.1 BERT trained on WikiPII dataset for 1 to 10 epochs

Epoch	Precision	Recall	F1-Score
1	0.91	0.94	0.93
2	0.93	0.95	0.94
3	**0.93**	**0.95**	**0.94**
4	0.93	0.95	0.94
5	0.94	0.95	0.94
6	0.93	0.95	0.94
7	0.93	0.94	0.94
8	0.92	0.94	0.93
9	0.94	0.95	0.94
10	0.93	0.94	0.93

A. V. D. M. Kayem, *Enabling Privacy Preserving Data Analytics*, Advances in Information Security 92, https://doi.org/10.1007/978-3-031-93906-8

Appendix B
Data Preprocessing and Transfer Learning Architecture

B.1 Data Pre-Processing

In the data pre-processing phase, the dataset was cleaned and edited to support PII detection. To do so, the data was annotated with labelled PII elements to enable training and testing. Different ratios of train-test splits of the data were also assessed to capture the diverse PII representations within the data (Fig. B.1).

To begin the annotation process the datasets were tokenised and named entity recognition (NER) was used to identify and classify selected entities based on a tagging scheme. A pre-trained transformer model was also use to support automated annotation [170]. To further support annotation, Regex (Regular Expressions) were employed on the datasets to label entities with specific (uniform) formats, such as phone numbers and email addresses [148]. Using such a hybrid strategy combining both automated methods, together with manual checking was aimed at ensuring we get optimal labelling results in the pre-processing phase. While this annotation process is more time-efficient, there is a potential risk of introducing noise into the

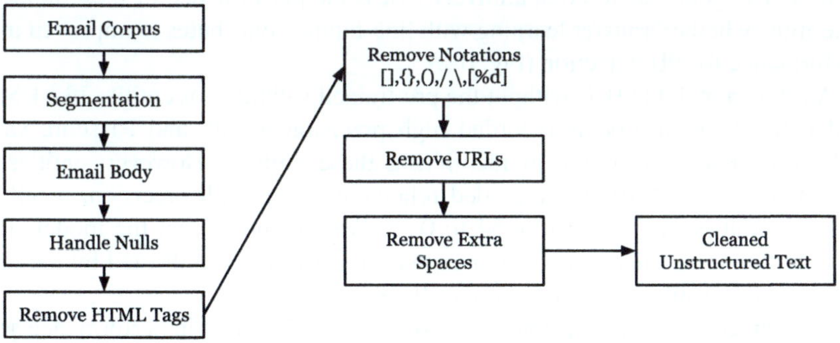

Fig. B.1 Data pre-processing schema [197, 198]

A. V. D. M. Kayem, *Enabling Privacy Preserving Data Analytics*, Advances in Information Security 92, https://doi.org/10.1007/978-3-031-93906-8

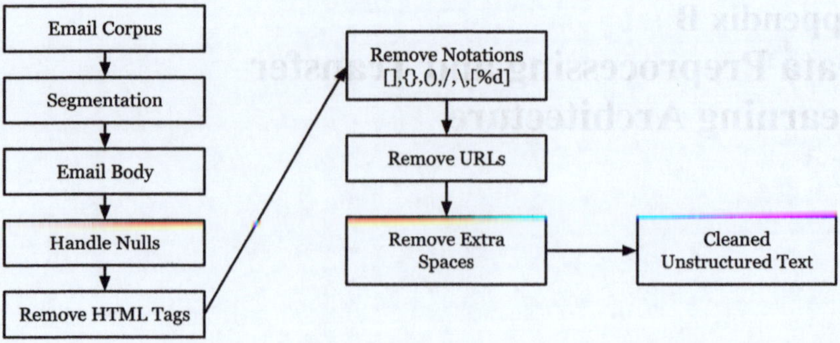

Fig. B.2 Data pre-processing schema [197, 198]

datasets. However, transformer models have been shown to be resilient to noise in data [181].

We tested various splits of data to ensure that the models were trained on a broad subset of the data with high variability to ensure generalisability during the testing phase as well as preventing overfitting [17].

B.2 Transfer Learning

Transfer learning exploits knowledge acquired from one task to improve the accuracy of results from subsequent tasks, and improves performance by assessing the relative gains from multiple candidate solutions. In our context, we looked at enhancing NER models using transfer learning techniques. To achieve this, we studied two BERT models namely—BERT and RoBERTa, which were trained. In each case, one version of the model was trained with pre-trained weights and the other without. The train-test split of the data was 70–30% the WikiPII dataset. The goal was to comparatively assess the performance of these models to determine whether transfer learning with fine-tuning contributes to improved model performance for PII detection (Fig. B.2).

As shown in Table B.1, without the pre-trained weights, for entities like EMAIL and PHONE, both models recorded high precision, recall, and F1-score values, indicating a high accuracy in identifying these entities. However, entities like LOCATION and PERSON, recorded relatively lower, with precision, recall, and F1-score values ranging from 0.54 to 0.65. We also noted that the models failed to identify entities that involved contextual information, as indicated by the macro average (across all entities) F1-score of 0.79.

Pre-trained weights, contribute to an overall improvement in performance across all entity types. The precision, recall, and F1-score are consistently higher and the precision result is 0.95 and the recall result 0.96, as highlighted in Table B.1. This indicates that pre-training BERT type models on contextual data plays a crucial

Table B.1 Effect of model performance using transfer learning

	Without pre-trained weights			With pre-trained weights		
Entity	Precision	Recall	F1-score	Precision	Recall	F1-score
Email	0.97	0.97	0.97	1.00	1.00	1.00
Location	0.58	0.51	0.54	0.89	0.91	0.90
Person	0.58	0.74	0.65	0.90	0.93	0.91
Phone	0.98	0.99	0.99	1.00	1.00	1.00
Macro avg.	0.78	0.80	0.79	**0.95**	**0.96**	**0.95**

role in enhancing its performance on NER tasks as demonstrated by the substantial improvement in detection when using pre-trained weights. It also sheds light why outlier detection failed to achieve very high precision in detecting PII.

Glossary

ACE	Automated Concatenation of Embeddings for Structured Prediction.
API	Application Programming Interface.
AUC	Area Under the Curve.
BERT	Bidirectional Encoder Representations from Transformers.
Bi-LSTM	Bidirectional Long Short-Term Memory.
BioNER	Biomedical Named Entity Recognition.
BILOU	(Beginning, Inside, Last, Outside, Unit) tagging scheme.
BIOES	(Beginning, Inside, Outside, End, Single) tagging scheme.
BoW	Bag-of-Words.
BPE	Byte-Pair Encoding.
CALO	A Cognitive Assistant that Learns and Organizes.
CBOW	Continuous Bag of Words.
CNN	Convolutional Neural Network.
CoNLL	Conference on Computational Natural Language Learning.
COPPA	Children's Online Privacy Protection Act.
CRF	Conditional Random Fields.
CUDA	Compute Unified Device Architecture.
DBSCAN	Density-Based Spatial Clustering of Applications with Noise.
ELMo	Embeddings from Language Models.
ER	Entity Ratio.
FCL	Fully-connected Layer.
FN	False Negative.
FNR	False Negative Rate.
FP	False Positive.
FPO	False Positives Labeled as 'O'.
FPR	False Positive Rate.
GDPR	General Data Protection Regulation.
GLBA	Gramm-Leach-Bliley Act.
GloVe	Global Vectors for Word Representation.

A. V. D. M. Kayem, *Enabling Privacy Preserving Data Analytics*, Advances in Information Security 92, https://doi.org/10.1007/978-3-031-93906-8

GPT	Generative Pre-Trained Transformers.
GPU	Graphics Processing Unit.
HIPAA	Health Insurance Portability and Accountability Act.
HMM	Hidden Markov Models.
HTML	HyperText Markup Language.
IDF	Inverse Document Frequency.
IEEE	Institute of Electrical and Electronics Engineers.
iForest	Isolation Forest.
IO	(Inside, Outside) tagging scheme.
IOB	(Inside, Outside Beginning) tagging scheme.
LOF	Local Outlier Factor.
LSTM	Long Short-Term Memory.
MLM	Masked Language Mode.
MUC-6	Sixth Message Understanding Conference.
MUC-7	Seventh Message Understanding Conference.
NER	Named Entity Recognition.
NII	Personal Non-identifying Information.
NLP	Natural Language Processing.
NSP	Next Sentence Prediction.
OCR	Optical Character Recognition.
OCSVM	One Class Support Vector Machine.
OOV	Out-of-vocabulary.
OSN	Online Social Networks.
PII	Personal Identifying Information.
PIPEDA	Personal Information Protection and Electronic Documents Act.
POS	Part-of-speech.
Regex	Regular Expressions.
RNN	Rrecurrent Neural Network.
ROC	Receiver Operating characteristic.
RoBERTa	Robustly Optimized BERT Pretraining Approach.
SD	Standard Deviation.
SUOD	Scalable Unsupervised Outlier Detection.
TF	Term Frequency.
TF-IDF	Term Frequency-Inverse Document Frequency.
TN	True Negative.
TP	True Positive.
TPR	True Positive Rate.
TTR	Type-Token Ratio.
URL	Uniform Resource Locator.
UTF-8	Unicode Transformation Format—8-bit.
VPPA	Video and Library Privacy Protection Act.
XML	Extensible Markup Language.

References

1. Abawajy, J. H., Ninggal, M. I. H., & Herawan, T. (2016). Privacy preserving social network data publication. In *IEEE Communications Surveys & Tutorials, 18*(3), 1974–1997 (2016). https://doi.org/10.1109/COMST.2016.2533668
2. Abedjan, Z., & Naumann, F. (2011). Advancing the discovery of unique column combinations. In *Proceedings of the 20th ACM international conference on information and knowledge management. CIKM 11. Glasgow, Scotland, UK: Association for Computing Machinery, 2011* (pp. 1565–1570). ISBN: 9781450307178. https://doi.org/10.1145/2063576.2063801.
3. Abuelgasim, A., & Kayem, A. (2016). An approach to personalized privacy policy recommendations on online social networks. In *Proceedings of the 2nd international conference on information systems security and privacy (ICISSP 2016)* (pp. 126–137). SCITEPRESS - Science and Technology Publications. ISBN: 978-989-758-167-0.
4. Abuelgasim, A., & Kayem, A. V. D. M. (2017). A snow-ball algorithm for automating privacy policy configuration in social media. In O. Camp, S. Furnell, & P. Mori. (Eds.). *Information systems security and privacy* (pp. 108–129). Springer International Publishing. ISBN: 978-3-319-54433-5.
5. Acar, G., et al. (2013). FPDetective: Dusting the web for fingerprinters. In *Proceedings of the 2013 ACM SIGSAC conference on computer and communications security. CCS '13. Berlin, Germany* (pp. 1129–1140). Association for Computing Machinery. ISBN: 9781450324779. https://doi.org/10.1145/2508859.2516674.
6. Acar, G., et al. (2014). The web never forgets: Persistent tracking mechanisms in the wild. In *Proceedings of the 2014 ACM SIGSAC conference on computer and communications security* (pp. 674–689).
7. Acock, A. C., & Stavig, G. R. (1979). A measure of association for nonparametric statistics. *Social Forces, 57*(4), 1381–1386. ISSN: 0037-7732. eprint: https://academic.oup.com/sf/article-pdf/57/4/1381/6889176/57-4-1381.pdf. https://doi.org/10.1093/sf/57.4.1381.
8. Acquisti, A., et al. (2017). Nudges for privacy and security: Understanding and assisting users choices online. In *ACM Computer Survey, 50*(3). ISSN: 0360-0300. https://doi.org/10.1145/3054926.
9. Adomavicius, G., & Yang, M. (2022). Integrating behavioral, economic, and technical insights to understand and address algorithmic bias: A human-centric perspective. In *ACM Transactions on Management Information Systems, 13*(3). ISSN: 2158-656X. https://doi.org/10.1145/3519420.

A. V. D. M. Kayem, *Enabling Privacy Preserving Data Analytics*, Advances in Information Security 92, https://doi.org/10.1007/978-3-031-93906-8

10. Aggarwal, C. C. (2005). On k-anonymity and the curse of dimensionality. In *Proceedings of the 31st international conference on very large data bases (VLDB '05). Trondheim, Norway* (pp. 901–909). VLDB Endowment.
11. Aggarwal, C. C. (2015). Outlier analysis. In *Data mining: The textbook* (pp. 237–263). Springer International Publishing. ISBN: 978-3-319-14142-8. https://doi.org/10.1007/978-3-319-14142-8_8.
12. Aggarwal, C. C. (2017). An introduction to outlier analysis. In *Outlier analysis* (pp. 1–34). Springer.
13. Aggarwal, C. C., & Sathe, S. (2015). Theoretical foundations and algorithms for outlier ensembles. *ACM SIGKDD Explorations Newsletter, 17*(1), 24–47.
14. Aggarwal, C. C., & Yu, P. S. (2008). A general survey of privacy-preserving data mining models and algorithms. In *Privacy-preserving data mining* (pp. 11–52). Springer.
15. Agrawal, R., et al. (2010). *Auditing compliance with a hippocratic database*. US Patent 7,810,142.
16. Agresti, A. (2012). *Categorical data analysis* (3rd ed.). John Wiley & Sons Inc.
17. Al-Riyami, S., Lisitsa, A., & Coenen, F. (2022). Cross-datasets evaluation of machine learning models for intrusion detection systems. In X. S. Yang et al. (Eds.),*Proceedings of sixth international congress on information and communication technology* (pp.815–828). Springer Singapore. ISBN: 978-981-16-2102-4.
18. Alsalibi, B., & Zakaria, N. (2013). CFPRS: Collaborative filtering privacy recommender system for online social networks. In *International Journal of Engineering research and application, 3*, 1850–1858.
19. An, S., et al. (2018). Contact graph based anonymization for geosocial network datasets. In *2018 5th international conference on behavioral, economic, and socio-cultural computing (BESC)* (pp. 132–137). https://doi.org/10.1109/BESC.2018.8697836.
20. Arasu, A., et al. (2014). Querying encrypted data. In *Proceedings of the 2014 ACM SIGMOD international conference on management of data (SIGMOD '14). Snowbird, Utah, USA* (pp. 1259–1261). ACM. ISBN: 978-1-4503-2376-5.
21. Ayenson, M. D., et al. (2011). Flash cookies and privacy II: Now with HTML5 and ETag respawning. Available at SSRN 1898390
22. Babaro, M., & Zeller, T., Jr. (2006). A face is exposed for AOL searcher No. 4417749. *The New York Times*. https://www.nytimes.com/2006/08/09/technology/09aol.html.
23. Backstrom, L., Dwork, C., & Kleinberg, J. (2007). Wherefore art thou R3579x? Anonymized social networks, hidden patterns, and structural steganography. In *Proceedings of the 16th international conference on world wide web (WWW '07), Banff, Alberta, Canada* (pp. 181–190). Association for Computing Machinery. ISBN: 9781595936547. https://doi.org/10.1145/1242572.1242598.
24. Barbaro, M., & Zeller, T. (2006). *A face is exposed for AOL searcher No. 4417749*. http://www.nytimes.com/2006/08/09/technology/09aol.html.
25. Barnett, V., & Lewis, T. (1994). Outliers in statistical data. *Probability, Mathematics, and Statistics* (3rd edn). Wiley.
26. Barth-Jones, D. (2012). The're-identification'of Governor William Weld's medical information: A critical re-examination of health data identification risks and privacy protections, then and now. In *Then and now (July 2012)*.
27. Bayardo, R. J., & Agrawal, R. (2005). Data privacy through optimal K-anonymization. In *21st international conference on data engineering (ICDE'05)* (pp. 217–228)
28. Beretta, S., et al. (2009). Parameterized complexity of k-anonymity: Hardness and tractability. Preprint, arXiv:0910.3148.
29. Birnick, J., et al. (2020). Hitting set enumeration with partial information for unique column combination discovery. *Proceedings of the VLDB Endowment, 13*(11), 2270–2283
30. Bläsius, T., Friedrich, T., & Schirneck, M. (2016). The parameterized complexity of dependency detection in relational databases. In *11th International Symposium on Parameterized and Exact Computation (IPEC 2016)*. Ed. by Jiong Guo and Danny Hermelin. Vol. 63. Leibniz International Proceedings in Informatics (LIPIcs). Dagstuhl, Germany: Schloss Dagstuhl–

Leibniz-Zentrum fuür Informatik, 2017, 6:1–6:13. ISBN: 978-3-95977-023-1. https://doi.org/10.4230/LIPIcs.IPEC.2016.6. http://drops.dagstuhl.de/opus/volltexte/2017/6920.
31. Blouvshtein, L., & Cohen-Or, D. (2019). Outlier detection for robust multi-dimensional scaling. *IEEE Transactions on Pattern Analysis and Machine Intelligence, 41*(9), 2273–2279.
32. Blum, A., et al. (2005). Practical privacy: The SuLQ framework. In *Proceedings of the twenty-fourth ACM SIGMOD-SIGACT-SIGART symposium on Principles of database systems* (pp. 128–138).
33. Boukerche, A., Zheng, L., & Alfandi, O. (2020). Outlier detection: Methods, models, and classification. In *ACM Computer Survey, 53*(3). ISSN: 0360-0300. https://doi.org/10.1145/3381028.
34. Boukoros, S., & Katzenbeisser, S. (2017). Measuring privacy in high dimensional microdata collections. In *Proceedings of the 12th international conference on availability, reliability and security. ARES '17. Reggio Calabria, Italy*. Association for Computing Machinery. ISBN: 9781450352574. https://doi.org/10.1145/3098954.3098977.
35. Brandaão, A., Mendes, R., & Vilela, J. P. (2022). Prediction of mobile app privacy preferences with user profiles via federated learning. In *Proceedings of the twelfth ACM conference on data and application security and privacy (CODASPY '22) Baltimore, MD, USA* (pp. 89–100). Association for Computing Machinery. ISBN: 9781450392204. https://doi.org/10.1145/3508398.3511526.
36. Bredereck, R., et al. (2014). The effect of homogeneity on the computational complexity of combinatorial data anonymization. In *Data Mining and Knowledge Discovery, 28*(1), 65–91 (2014)
37. Breunig, M. M., et al. (2000). LOF: Identifying density-based local outliers. In *Proceedings of the 2000 ACM SIGMOD international conference on management of data (SIGMOD '00)* (pp. 93–104). Association for Computing Machinery.
38. Budak, C., et al. (2016). Understanding emerging threats to online advertising. In *Proceedings of the 2016 ACM conference on economics and computation* (pp. 561–578).
39. Burke, M.-J., & Kayem, A. V. D. M. (2014). *k*-anonymity for privacy preserving crime data publishing in resource constrained environments. In *28th international conference on advanced information networking and applications workshops, AINA 2014 Workshops, Victoria, BC, Canada, May 13–16, 2014* (pp. 833–840).
40. Cerf, V. G. (2022). Preserving the internet. *Communications of the ACM, 65*(4), 5. https://doi.org/10.1145/3522782. ISSN: 0001-0782.
41. Chakraborty, D., Narayanan, V., & Ghosh, A. (2019). Integration of deep feature extraction and ensemble learning for outlier detection. *Pattern Recognition, 89*, 161–171.
42. Chakraborty, S., & Tripathy, B. K. (2015). Privacy preservation in social networks through alpha Anonymization techniques. In *2015 IEEE/ACM international conference on advances in social networks analysis and mining (ASONAM)* (pp. 1602–1603). https://doi.org/10.1145/2808797.2809354.
43. Chawla, S., et al. (2005). Toward privacy in public databases. In *Theory of, cryptography conference* (pp. 363–385). Springer.
44. Chuang, I-H., et al. (2011). An effective privacy protection scheme for cloud computing. In *13th international conference on advanced communication technology (ICACT2011)* (pp. 260–265).
45. Ciglic, M., Eder, J., & Koncilia, C. (2014). *k*-anonymity of microdata with NULL values. In *Database and Expert Systems Applications - 25th International Conference, DEXA 2016, Munich, Germany, September 2014, Proceedings, Part I* (pp. 328–342).
46. Ciglic, M., Eder, J., & Koncilia, C. (2016). Anonymization of data sets with NULL values. In A. Hameurlain, et al. (Eds.), *Transactions on large-scale data- and knowledge-centered systems XXIV: Special issue on database- and expert-systems applications* (pp. 193–220). Springer Berlin Heidelberg. ISBN: 978-3-662-49214-7. https://doi.org/10.1007/978-3-662-49214-7_7.
47. Cohen, W. (2015). *Enron email dataset*. Carnegie Mellon University. Accessed Mar 08, 2023, https://www.cs.cmu.edu/enron.

48. European Commission. (2019). *General data protection regulation*. https://ec.europa.eu/info/law/law-topic/data-protection/reform/what-does-general-data-protection-regulation-gdpr-govern_en.
49. Cormode, G., et al. (2010). Minimizing minimality and maximizing utility: Analyzing method-based attacks on anonymized data. In *Proceeedng of the VLDB Endowment, 3*(1–2), 1045–1056. https://doi.org/10.14778/1920841.1920972. ISSN: 2150-8097.
50. Cormode, G., et al. (2018). Privacy at scale: Local differential privacy in practice. In *Proceedings of the 2018 international conference on management of data (SIGMOD '18) Houston, TX, USA* (pp. 1655–1658). Association for Computing Machinery. ISBN: 9781450347037. https://doi.org/10.1145/3183713.3197390.
51. Cyber-Awareness. (2019). *Breach level index - Calculate your breach level index*. https://kybersecure.com/cyber-breach-level-index/ (visited on Oct 10, 2022).
52. Dalenius, T. (1986). Finding a needle in a haystack - or identifying anonymous census record. *Journal of Official Statistics, 2*(3), 329–336.
53. Dankar, F. K., & Emam, K. E. (2012). The application of differential privacy to health data. In *Proceedings of the 2012 joint EDBT/ICDT workshops* (pp. 158–166).
54. Dankar, F. K., & Emam, K. E. (2013). Practicing differential privacy in health care: A review. *Transactions on Data Privacy, 6*(1), 35–67 (2013).
55. Das, S., et al. (2016). Incorporating expert feedback into active anomaly discovery. In *2016 IEEE 16th international conference on data mining (ICDM)* (pp. 853–858).
56. Davis, J. (2020). *Health data, medical documents exposed by labcorp website error*. https://healthitsecurity.com/news/health-data-medical-documents-exposed-by-labcorp-website-error.
57. Denning, D. E., & Denning, P. J. (1979). The tracker: A threat to statistical database security. *ACM Transactions on Database Systems (TODS), 4*(1), 76–96.
58. Deésir, C., et al. (2013). One class random forests. In *Pattern Recognition, 46*(12), 3490–3506. https://doi.org/10.1016/j.patcog.2013.05.022. ISSN: 0031-3203.
59. Desolda, G., et al. (2022). Human factors in phishing attacks: A systematic literature review. *ACM Computing Surveys (CSUR), 54*, 1–35.
60. Ding, X., et al. (2011). De-anonymizing dynamic social networks. In *2011 IEEE global telecommunications conference - GLOBECOM 2011* (pp. 1–6). https://doi.org/10.1109/GLOCOM.2011.6133607.
61. Domingo-Ferrer, J., Sanchez, D., & Soria-Comas, J. (2016). *Database anonymization: privacy models, data utility, and microaggregation-based inter-model connections* (Vol. 8, pp. 1–136). Morgan & Claypool Publishers.
62. Dondi, R., Mauri, G., & Zoppis, I. (2011). On the complexity of the l-diversity problem. In *MFCS* (pp. 266–277). Springer.
63. Duan, L., et al. (2009). Cluster-based outlier detection. *Annals of Operations Research, 168*(1), 151–168 (2009)
64. Dwork, C. (2008). Differential privacy: A survey of results. In *International conference on theory and applications of models of computation* (pp. 1–19). Springer.
65. Dwork, C. (2011). Differential privacy. In *Encyclopedia of cryptography and security* (pp. 338–340). Springer.
66. Dwork, C., & Roth, A. (2014). The algorithmic foundations of differential privacy. *Foundations and Trends in Theoretical Computer Science, 9*(3–4), 211–407.
67. Dwork, C., et al. (2006). Calibrating noise to sensitivity in private data analysis. In *Theory of cryptography conference* (pp. 265–284). Springer.
68. Dwork, C., et al. (2009). On the complexity of differentially private data release: Efficient algorithms and hardness results. In *Proceedings of the forty-first annual ACM symposium on theory of computing (STOC '09). Bethesda, MD, USA* (pp. 381–390). ACM. ISBN: 978-1-60558-506-2. http://doi.acm.org/10.1145/1536414.1536467.
69. Dyer, C., et al. (2015). Transition-based dependency parsing with stack long short-term memory. Preprint, arXiv:1505.08075.

70. Emam, K. E. (2011). Methods for the de-identification of electronic health records for genomic research. *Genome Medicine, 3*(4), 25.
71. Englehardt, S., & Narayanan, A. (2016). Online tracking: A 1-million- site measurement and analysis. In *Proceedings of the 2016 ACM SIGSAC conference on computer and communications security* (pp. 1388–1401).
72. Englehardt, S., et al. (2015). Cookies that give you away: The surveillance implications of web tracking. In *Proceedings of the 24th international conference on world wide web* (pp. 289–299).
73. El-Yahyaoui, A., & Kettani, M. D. E-C. E. (2017). Fully homomorphic encryption: Searching over encrypted cloud data. In *Proceedings of the 2nd international conference on big data, cloud and applications (BDCA'17) Tetouan, Morocco*. Association for Computing Machinery. ISBN: 9781450348522. https://doi.org/10.1145/3090354.3090364.
74. eSito. (2021). *eSito: g9 anonymizer*. https://www.esito.no/en/products/anonymizer/ (visited on Apr 10, 2021).
75. Everitt, B. S., & Hothorn, T. (2009). *A handbook of statistical analyses using r* (2nd ed.). Chapman & Hall/CRC. ISBN: 1420079336.
76. Evfimievski, A., Gehrke, J., & Srikant, R. (2003). Limiting privacy breaches in privacy preserving data mining. In *Proceedings of the twenty-second ACM SIGMOD-SIGACT-SIGART symposium on principles of database systems (PODS '03). San Diego, California*(pp. 211–222). Associa-tion for Computing Machinery. ISBN: 1581136706. https://doi.org/10.1145/773153.773174.
77. Fang, L., & LeFevre, K. (2010). Privacy wizards for social networking sites. In *Proceedings of the 19th international conference on world wide web (WWW '10). Raleigh, North Carolina, USA* (pp. 351–360). Association for Computing Machinery. ISBN: 9781605587998. https://doi.org/10.1145/1772690.1772727.
78. Fienberg, S. E., & Jin, J. (2012). Privacy-preserving data sharing in high dimensional regression and classification settings. *Journal of Privacy and Confidentiality, 4*(1), 10.
79. Fire, M., Goldschmidt, R., & Elovici, Y. (2014). Online social networks: Threats and solutions. *IEEE Communications Surveys Tutorials, 16*(4), 2019–2036. https://doi.org/10.1109/COMST.2014.2321628.
80. Frik, A., et al. (2022). Users expectations about and use of smartphone privacy and security settings. In *Proceedings of the 2022 CHI conference on human factors in computing systems (CHI '22). New Orleans, LA, USA*. Association for Computing Machinery. ISBN: 9781450391573. https://doi.org/10.1145/3491102.3517504.
81. Fu, L., et al. (2020). De-anonymizing social networks with overlapping community structure. *IEEE/ACM Transactions on Networking, 28*(1), 360–375. https://doi.org/10.1109/TNET.2019.2962731.
82. Fung, B. C. M., et al. (2010). Privacy-preserving data publishing: A survey of recent developments. In *ACM Computer Survey, 42*(4). ISSN: 0360-0300. https://doi.org/10.1145/1749603.1749605.
83. Gandomi, A., & Haider, M. (2015). Beyond the hype: Big data concepts, methods, and analytics. *International Journal of Information Management, 35*(2), 137–144.
84. Ganta, S. R., Kasiviswanathan, S. P., & Smith, A. (2008). Composition attacks and auxiliary information in data privacy. In *Proceedings of the 14th ACM SIGKDD international conference on knowledge discovery and data mining (KDD '08). Las Vegas, Nevada, USA* (pp. 265–273). Association for Computing Machinery. ISBN: 9781605581934. https://doi.org/10.1145/1401890.1401926.
85. Gao, T., & Li., F. (2019). Sharing social networks using a novel differentially private graph model. In *2019 16th IEEE annual consumer communications & networking conference (CCNC)* (pp. 1–4). https://doi.org/10.1109/CCNC.2019.8651689.
86. Gao, T., & Li, F. (2022). De-anonymizing online social network with conditional generative adversarial network. In *2022 IEEE 19th international conference on mobile ad hoc and smart systems (MASS)* (pp. 496–504). https://doi.org/10.1109/MASS56207.2022.00076.

87. Gao, H., et al. (2011). Security issues in online social networks. *IEEE Internet Computing, 15*(4), 56–63. https://doi.org/10.1109/MIC.2011.50.
88. García-Teodoroa, P., et al. (2009). Anomaly-based network intrusion detection: Tech-niques, systems and challenges. *Computers and Security, 28*(1), 18–28. ISSN: 0167-4048. https://doi.org/10.1016/j.cose.2008.08.003. https://www.sciencedirect.com/science/article/pii/S0167404808000692.
89. Gedik, B., & Liu, L. (2008). Protecting location privacy with personalized k-anonymity: Architecture and algorithms. *IEEE Transactions on Mobile Computing, 7*(1), 1–18. https://doi.org/10.1109/TMC.2007.1062.
90. Ghazinour, K., Matwin, S., & Sokolova, M. (2013). Monitoring and recommending privacy settings in social networks. In *Proceedings of the joint EDBT/ICDT 2013 workshops (EDBT '13) Genoa, Italy* (pp. 164–168). Association for Computing Machinery. https://doi.org/10.1145/2457317.2457344. ISBN: 9781450315999.
91. Ghazinour, K., Matwin, S., & Sokolova, M. (2013). YourPrivacyProtector: A recommender system for privacy settings in social networks. *International Journal of Security, Privacy and Trust Management, 2*, 11–25. https://doi.org/10.5121/ijsptm.2013.2402.
92. Giovannetti, E., & Hamoudia, M. (2018). *Understanding the different pre and post peak adoption drivers in the process of Mobile Social Network-ing diffusion* (pp. 1–4). International Telecommunications Society (ITS).
93. Golle, P. (2006). Revisiting the uniqueness of simple demographics in the US population. In *Proceedings of the 5th ACM workshop on privacy in electronic society (WPES '06). Alexandria, Virginia, USA* (pp. 77–80). Association for Computing Machinery. ISBN: 1595935568. https://doi.org/10.1145/1179601.1179615.
94. Gross, R., & Acquisti, A. (2005). Information revelation and privacy in online social networks. In *Proceedings of the 2005 ACM workshop on privacy in the electronic society (WPES '05) Alexandria, VA, USA* (pp. 71–80). Association for Computing Machinery. ISBN: 1595932283. https://doi.org/10.1145/1102199.1102214.
95. Gundecha, P., Barbier, G., & Liu, H. (2011). Exploiting vulnerability to secure user privacy on a social networking site. In *Proceedings of the 17th ACM SIGKDD international conference on knowledge discovery and data mining, KDD'11*. Sept. 2011, pp. 511–519. ISBN: 978-145-030-813-7. https://doi.org/10.1145/2020408.2020489.
96. Hall, M., et al. (2009). The WEKA data mining software: An update. *SIGKDD Explorations Newsletter, 11*(1), 10–18. https://doi.org/10.1145/1656274.1656278. ISSN: 1931-0145.
97. Hansell, S. (2006). AOL removes search data on group of web users. *The New York Times*. https://www.nytimes.com/2006/08/08/business/media/08aol.html.
98. Hathurusinghe, R. (2020). Building a personally identifiable information recognizer in a privacy preserved manner using automated annotation and federated learning. https://doi.org/10.20381/ruor-25235. http://hdl.handle.net/10393/41011.
99. Hautamaki, V., Karkkainen, I., & Franti, P. (2004). Outlier detection using k-nearest neighbour graph. In *Proceedings of the 17th international con-ference on pattern recognition, 2004 (ICPR 2004)* (Vol. 3, pp. 430–433).
100. He, X., et al. (2014). Relevant overlapping subspace clusters on categorical data. In *Proceedings of the 20th ACM SIGKDD international conference on knowledge discovery and data mining (KDD '14). New York, New York, USA* (pp. 213–222). Association for Computing Machinery. ISBN: 9781450329569. https://doi.org/10.1145/2623330.2623652.
101. Heinrich, A., Bittner, N., & Hollick, M. (2022). AirGuard - protecting android users from stalking attacks by apple find my devices. In *Proceedings of the 15th ACM conference on security and privacy in wireless and mobile networks (WiSec '22) San Antonio, TX, USA* (pp. 26–38). Asso ciation for Computing Machinery. ISBN: 9781450392167. https://doi.org/10.1145/3507657.3528546.
102. Hipp, J., Guüntzer, U., & Nakhaeizadeh, G. (2000). Algorithms for association rule mining: A general survey and comparison. *SIGKDD Explorations Newsletter, 2*(1), 58–64. https://doi.org/10.1145/360402.360421. ISSN: 1931-0145.

103. Islam, M. R. , Kayem, A. V. D. M., & Meinel, C. (2023). Enabling PII discovery in textual data via outlier detection. In *Proceedings of the 34th international conference on database and expert systems applications (DEXA 2023)*. Springer (To Appear).
104. Johnson, A. E. W., Bulgarelli, L., & Pollard, T. J. (2020). Deiden-tification of free-text medical records using pre-trained bidirectional transformers. In *Proceedings of the ACM conference on health, inference, and learning (CHIL '20). Toronto, Ontario, Canada* (pp. 214–221). Association for Computing Machinery. ISBN: 9781450370462. https://doi.org/10.1145/3368555.3384455.
105. Joy, A. P., et al. (2021). Precise estimation of local probabilities for Bayesian attack graph analysis. In *2021 IEEE 12th annual information technology, electronics and mobile communication conference (IEMCON)* (pp. 0080–0085). https://doi.org/10.1109/IEMCON53756.2021.9623254.
106. Kantarcioglu, M., & Clifton, C. (2004). Privacy-preserving distributed mining of association rules on horizontally partitioned data. *IEEE Transactions on Knowledge and Data Engineering, 16*(9), 1026–1037.
107. Kayem, A. V. D. M., Vester, C. T., & Meinel, C. (2016). Automated k-Anonymization and l-Diversity for shared data privacy. In *Database and expert systems applications - 27th international conference, DEXA 2016, Porto, Portugal, September 5-8, 2016, Proceedings, Part I* (pp. 105–120). https://doi.org/10.1007/978-3-319-44403-1_7.
108. Kayem, A. V. D. M., Vester, C. T., & Meinel, C. (2018). Syntactic anonymisation of shared datasets in resource constrained environments. In *Transactions on Large-Scale Data- and Knowledge Centered System, 38*, 27–60. https://doi.org/10.1007/978-3-662-58384-5%5C_2.
109. Kifer, D., & Machanavajjhala, A. (2011). No free lunch in data privacy. In *Proceedings of the 2011 ACM SIGMOD international conference on management of data, SIGMOD '11. New York, NY, USA* (pp. 193–204). ACM. ISBN: 978-1-4503-0661-4. http://doi.acm.org/10.1145/1989323.1989345.
110. Kleinberg, B., et al. (2017). NETANOS-named entity-based text anonymization for open science. PsyArxiv Preprints.
111. Knorr, E. M., & Ng, R. T. (1998). Algorithms for mining distance- based outliers in large datasets. In *Proceedings of the 24rd international conference on very large data bases. VLDB '98* (pp. 392–403).
112. Kokkula, S., & Musti, N. M. (2013). Classification and outlier detection based on topic based pattern synthesis. In *Proceedings of the international workshop on machine learning and data mining in pattern recognition* (pp. 99–114). Springer.
113. Kriegel, H.-P., et al. (2009). Outlier detection in axis-parallel subspaces of high dimensional data. In *Pacific-Asia conference on knowledge discovery and data mining* (pp. 831–838). Springer.
114. Kumar, S., & Thilagam, S. (2019). Approaches and challenges of privacy preserving search over encrypted data. *Information Systems, 81*, 63–81. ISSN: 0306-4379. https://doi.org/10.1016/j.is.2018.11.004. https://www.sciencedirect.com/science/article/pii/S0306437917303824.
115. Kushida, C. A., et al. (2012). Strategies for de-identification and anonymization of electronic health record data for use in multicenter research studies. *Medical Care*, S82–S101.
116. Lample, G., et al. (2016). Neural architectures for named entity recognition. In *Proceedings of the 2016 conference of the North American chapter of the association for computational linguistics: Human language technologies* (pp. 260–270). Association for Computational Linguistics.
117. Last, M., et al. (2014). Improving accuracy of classification models induced from anonymized datasets. *Information Science, 256*, 138–161. https://doi.org/10.1016/j.ins.2013.07.034. ISSN: 0020-0255.
118. LeFevre, K., DeWitt, D. J., & Ramakrishnan, R. (2005). Incognito: Efficient full-domain *k*-anonymity. In *Proceedings of the 2005 ACM SIG- MOD international conference on management of data, SIGMOD '05. Baltimore, Maryland* (pp. 49–60). ACM. ISBN: 1-59593-060-4. http://doi.acm.org/10.1145/1066157.1066164.

119. LeFevre, K., DeWitt, D. J., & Ramakrishnan, R. (2006). Mondrian multidimensional K-anonymity. In *22nd international conference on data engineering (ICDE'06)* (pp. 25–25). https://doi.org/10.1109/ICDE.2006.101.
120. Lehmann, A. (2019). ScrambleDB: Oblivious (Chameleon) pseudonymization- as-a-service. *PoPETs, 3*, 289–309. https://doi.org/10.2478/popets-2019-0048.
121. Lenz, S. T., & Agresti, A. (2016). Categorical data analysis. *Statistical Papers*, 849–850. https://doi.org/10.1007/s00362-015-0733-8.
122. Li, H., et al. (2020). Privacy leakage via de-anonymization and aggregation in heterogeneous social networks. *IEEE Transactions on Dependable and Secure Computing, 17*(2), 350–362. https://doi.org/10.1109/TDSC.2017.2754249.
123. Li, N., Li, T., & Venkatasubramanian, S. (2007). t-Closeness: Privacy beyond k-anonymity and l-diversity. In *2007 IEEE 23rd international conference on data engineering* (pp. 106–115). https://doi.org/10.1109/ICDE.2007.367856.
124. Li, N., Li, T., & Venkatasubramanian, S. (2010). Closeness: A new privacy measure for data publishing. *IEEE Transactions on Knowledge and Data Engineering, 22*(7), 943–956. https://doi.org/10.1109/TKDE.2009.139.
125. Li, H., Zhang, Q., & Kuldeep, G. (2023). Demo: 3TierView - A three-tier privacy-preserving live video surveillance IoT system. In *Proceedings of the 21st annual international conference on mobile systems, applications and services (MobiSys '23) Helsinki, Finland* (pp. 594–595). Association for Computing Machinery. https://doi.org/10.1145/3581791.3597289.
126. Liang, H., & Yuan, H. (2013). On the complexity of t-closeness anonymization and related problems. In W. Meng, et al. (Eds.), *Database systems for advanced applications* (pp. 331–345). Springer Berlin Heidelberg. ISBN: 978-3-642-37487-6.
127. Liang, K. et al. (2016). Efficient multi-function data sharing and searching mechanism for cloud-based encrypted data. In *Proceedings of the 11th ACM on Asia conference on computer and communications security (ASIA CCS '16). Xi'an, China* (83–94). Association for Computing Machinery. ISBN: 9781450342339. https://doi.org/10.1145/2897845.2897865.
128. Libert, T. (2015). Exposing the hidden web: An analysis of third-party HTTP requests on 1 million websites. *International Journal of Communication*. https://ssrn.com/abstract=2685330/.
129. Liu, K., Kargupta, H., & Ryan, J. (2006). Random projection-based multiplicative data perturbation for privacy preserving distributed data mining. *IEEE Transactions on knowledge and Data Engineering, 18*(1), 92–106.
130. Liu, Y., et al. (2011). Analyzing facebook privacy settings: User expectations vs. reality. In *Proceedings of the 2011 ACM SIGCOMM conference on internet measurement conference (IMC '11). Berlin, Germany* (pp. 61–70). Association for Computing Machinery. ISBN: 9781450310130. https://doi.org/10.1145/2068816.2068823.
131. Liu, X., et al. (2012). Automatic taxonomy construction from keywords. In *Proceedings of the 18th ACM SIGKDD international conference on knowledge discovery and data mining (KDD '12). Beijing, China* (pp. 1433–1441). Association for Computing Machinery. ISBN: 9781450314626. https://doi.org/10.1145/2339530.2339754.
132. Liu, Y., et al. (2015). Computing semantic text similarity using rich features. In *Proceedings of the 29th Pacific Asia conference on language, information and computation* (pp. 44–52).
133. Liu, H., et al. (2018). Efficient outlier detection for high-dimensional data. *IEEE Transactions on Systems, Man, and Cybernetics: Systems, 48*(12), 2451–2461.
134. Liu, Y., et al. (2020). Generative adversarial active learning for unsupervised outlier detection. *IEEE Transactions on Knowledge and Data Engineering, 32*(8), 1517–1528.
135. Liu, C., et al. (2022). Privacy intelligence: A survey on image privacy in online social networks. *ACM Computing Surveys*. ISSN: 0360-0300. https://doi.org/10.1145/3547299.
136. Liu, P., et al. (2022). A survey on active deep learning: from model driven to data driven. *ACM Computer Surveys, 54*(10s). ISSN: 0360-0300. https://doi.org/10.1145/3510414.
137. Ma, C. Y. T., et al. (2010). Privacy vulnerability of published anonymous mobility traces. In *Proceedings of the sixteenth annual international conference on Mobile computing and networking* (pp. 185–196).

138. Ma, J., et al. (2018). De-anonymizing social networks with random forest classifier. *IEEE Access, 6*, 10139–10150. https://doi.org/10.1109/ACCESS.2017.2756904.
139. Machanavajjhala, A., et al. (2007). L-diversity: Privacy beyond k-anonymity. *ACM Transactions on Knowledge Discovery from Data, 1*(1), 3–55 (3–es). ISSN: 1556-4681.
140. Madejski, M., Johnson, M., & Bellovin, S. M. (2012). A study of privacy settings errors in an online social network. In *2012 IEEE international conference on pervasive computing and communications workshops* (pp. 340–345). https://doi.org/10.1109/PerComW.2012.6197507.
141. Mainardi, N., Barenghi, A., & Gerardo, P. (2021). Privacy-aware character pattern matching over outsourced encrypted data. *Digital Threats, 3*(1). ISSN: 2692-1626. https://doi.org/10.1145/3462333.
142. Martin, N. (2019). *How much data is collected every minute of the day*. https://www.forbes.com/sites/nicolemartin1/2019/08/07/how-much-data-is-collected-every-minute-of-the-day/?sh=4602f8103d66 (visited on Oct 10, 2022).
143. Martin, D. J., et al. (2007). Worst-case background knowledge for privacy-preserving data publishing. In *2007 IEEE 23rd international conference on data engineering* (pp. 126–135). IEEE.
144. Meyerson, A., & Williams, R. (2004). On the complexity of optimal k- anonymity. In *Proceedings of the twenty-third ACM SIGMOD-SIGACT- SIGART symposium on principles of database systems (PODS '04). Paris, France* (pp. 223–228). Association for Computing Machinery. ISBN: 158113858X. https://doi.org/10.1145/1055558.1055591.
145. Migliarini, P., et al. (2020). On the elicitation of privacy and ethics preferences of mobile users. In *Proceedings of the IEEE/ACM 7th international conference on mobile software engineering and systems (MOBILESoft '20) Seoul, Republic of Korea* (pp. 132–136). Association for Computing Machinery. ISBN: 9781450379595. https://doi.org/10.1145/3387905.3388610.
146. Miklau, G., & Suciu, D. (2007). A formal analysis of information disclosure in data exchange. *Journal of Computer and System Sciences, 73*(3), 507–534.
147. Mirsky, Y., & Lee, W. (2021). The creation and detection of deepfakes: A survey. *ACM Computing Surveys, 54*(1), 1–41. https://doi.org/10.1145/3425780. ISSN: 0360-0300.
148. Mitra, M., & Roy, S. (2018). Identification and processing of pii data, applying deep learning models with improved accuracy and efficiency. *Journal of Data Acquisition and Processing, 34*(6), 1339–1340. https://sjcjycl.cn/article/view-2018/2018-1337.php.
149. Mondal, M., et al. (2019). Moving beyond set-it-and-forget-it privacy settings on social media. In *Proceedings of the 2019 ACM SIGSAC conference on computer and communications security (CCS '19). London, United Kingdom* (pp. 991–1008). Association for Computing Machinery. ISBN: 9781450367479. https://doi.org/10.1145/3319535.3354202.
150. de Montjoye, Y.-A., et al. (2013). Unique in the crowd: The privacy bounds of human mobility. *Scientific Reports, 3*, 1376.
151. de Montjoye, Y.-A., et al. (2015). Unique in the shopping mall: On the re- identifiability of credit card metadata. *Science, 347*(6221), 536–539. https://doi.org/10.1126/science.1256297. https://www.science.org/doi/abs/10.1126/science.1256297.
152. Motlagh, F. N., Kayem, A. V. D. M., & Meinel, C. (2022). A multi-agent model to support privacy preserving co-owned image sharing on social media. In L. Barolli, F. Hussain, & T. Enokido. (Eds.), *Advanced information networking and applications - proceedings of the 36th international conference on advanced information networking and applications (AINA-2022), Sydney, NSW, Australia, 13–15 April 2022, Volume 1*. Lecture Notes in Networks and Systems (Vol. 449, pp. 140–151). Springer. https://doi.org/10.1007/978-3-030-99584-3_13. https://doi.org/10.1007/978-3-030-99584-3%5C_13.
153. Mrabet, A., Bentounsi, M., & Darmon, P. (2019). SecP2I A secure multi-party discovery of personally identifiable information (PII) in structured and semi-structured datasets. In *2019 IEEE international conference on big data* (pp. 5028–5033). https://doi.org/10.1109/BigData47090.2019.9006605.

154. Munoz-Gonzalez, L., et al. (2019). Exact inference techniques for the analysis of Bayesian attack graphs. *IEEE Transactions on Dependable and Secure Computing, 16*(2), 231–244. https://doi.org/10.1109/TDSC.2016.2627033.
155. Nadeau, D., & Sekine, S. (2007). A survey of named entity recognition and classification. *Lingvisticae Investigationes, 30*(1), 3–26.
156. Narayanan, A., & Shmatikov, V. (2006). How to break anonymity of the netflix prize dataset. CoRR abs/cs/0610105. http://arxiv.org/abs/cs/0610105.
157. Narayanan, A., & Shmatikov, V. (2008). Robust de-anonymization of large sparse datasets. In *2008 IEEE symposium on security and privacy (sp 2008)* (pp. 111–125). IEEE.
158. Narayanan, A., & Shmatikov, V. (2009). De-anonymizing social networks. In *2009 30th IEEE symposium on security and privacy* (pp. 173–187). https://doi.org/10.1109/SP.2009.22.
159. Narayanan, A., et al. (2012). On the feasibility of internet-scale author identification. In *2012 IEEE symposium on security and privacy* (pp. 300–314). IEEE.
160. Neamatullah, I., et al. (2008). Automated de-identification of free-text medical records. *BMC Medical Informatics and Decision Making, 32*(8).
161. Nergiz, M. E., Tamersoy, A., & Saygin, Y. (2011). Instant anonymization. *ACM Transactions on Database Systems, 36*(1), 2:1– 2:33.
162. Niksirat, K. S., et al. (2024). Wearable activity trackers: A survey on utility, privacy, and security. In *ACM Computer Survey, 56*(7). ISSN: 0360-0300. https://doi.org/10.1145/3645091.
163. Ohm, P. (2009). Broken promises of privacy: Responding to the surprising failure of anonymization. *UCLA Law Review, 57*, 1701.
164. Ohrn, A., & Ohno-Machado, L. (1999). Using Boolean reasoning to anonymize databases. *Artificial Intelligence in Medicine, 15*(3), 235–254. ISSN: 0933-3657. https://doi.org/10.1016/S0933-3657(98)00056-6. https://www.sciencedirect.com/science/article/pii/S0933365798000566.
165. Olejnik, L., Castelluccia, C., & Janc, A. (2012). Why johnny can't browse in peace: On the uniqueness of web browsing history patterns. In *5th workshop on hot topics in privacy enhancing technologies (HotPETs 2012)*.
166. Ortloff, A.-M., et al. (2021). The effect of nudges and boosts on browsing privacy in a naturalistic environment. In *Proceedings of the 2021 conference on human information interaction and retrieval (CHIIR '21) Canberra ACT, Australia* (pp. 63–73). Association for Computing Machinery. ISBN: 9781450380553. https://doi.org/10.1145/3406522.3446014.
167. Oukemeni, S., Rifa'-Pous, H., & Puig, J. M. M. (2019). Privacy analysis on microblogging online social networks: A survey. *ACM Computer Survey, 52*(3). ISSN: 0360-0300. https://doi.org/10.1145/3321481.
168. Pang, G., et al. (2021). Deep learning for anomaly detection: A review. *ACM Computer Survey, 54*(2). ISSN: 0360-0300. https://doi.org/10.1145/3439950.
169. Papenbrock, T., & Naumann, F. (2017). A hybrid approach for efficient unique column combination discovery. In *Proceedings of the der Fachtagung business, technologie und web (BTW). GI, Bonn, Deutschland (accepted) Google Scholar* .
170. Pearson, C., Seliya, N., & Dave, R. (2021). Named entity recognition in unstructured medical text documents. In *2021 international conference on electrical, computer and energy technologies (ICECET)* (pp. 1–6). https://doi.org/10.1109/ICECET52533.2021.9698694.
171. Podlesny, N. J., et al. (2018). Minimising information loss on anonymised high dimensional data with greedy in-memory processing. In S. Hartmann, et al. (Eds.), *Database and expert systems applications - 29th international conference, DEXA 2018, Regensburg, Germany, September 3–6, 2018, proceedings, part I*. Lecture notes in computer science (Vol. 11029, pp. 85–100). Springer. https://doi.org/10.1007/978-3-319-98809-2_6. https://doi.org/10.1007/978-3-319-98809-2%5C_6.
172. Podlesny, N. J., Kayem, A. V. D. M., & Meinel, C. (2019). Attribute compartmentation and greedy UCC discovery for high-dimensional data anonymization. In G.-J. Ahn et al. (Eds.), *Proceedings of the ninth ACM conference on data and application security and privacy,*

CODASPY 2019, Richardson, TX, USA, March 25–27, 2019 (pp. 109–119). ACM. https://doi.org/10.1145/3292006.3300019.

173. Podlesny, N. J., Kayem, A. V. D. M., & Meinel, C. (2019). Towards identifying de-anonymisation risks in distributed health data silos. In S. Hartmann, et al. (Eds.), *Database and expert systems applications - 30th international conference, DEXA 2019, Linz, Austria, August 26–29, 2019, proceedings, part I*. Lecture notes in computer science (Vol. 11706, pp. 33–43). Springer. https://doi.org/10.1007/978-3-030-27615-7%5C_3.
174. Podlesny, N. J., et al. (2019). How data anonymisation techniques influence disease triage in digital health: A study on base rate neglect. In P. Kostkova, et al. (Eds.), *Proceedings of the 9th international conference on digital public health, PDH 2019, Marseille, France, November 20–23, 2019* (pp. 55–62). ACM. https://doi.org/10.1145/3357729.3357737.
175. Podlesny, N. J., Kayem, A. V. D. M., & Meinel, C. (2019). High-dimensional health data silos through bayesian networks optimised by multigrid and manifold. In *2019 IEEE international conference on dependable, autonomic and secure computing, international conference on pervasive intelligence and computing, international conference on cloud and big data computing, international conference on cyber science and technology congress, DASC/PiCom/CBDCom/CyberSciTech 2019, Fukuoka, Japan, August 5-8, 2019* (pp. 556–563). https://doi.org/10.1109/DASC/PiCom/CBDCom/CyberSciTech.2019.00110.
176. Podlesny, N. J., Kayem, A. V. D. M., & Meinel, C. (2021). A parallel quasi-identifier discovery scheme for dependable data anonymi-sation. *Transactions on Large-Scale Data- and Knowledge-Centered Systems, 50*, 1–24. https://doi.org/10.1007/978-3-662-64553-6_1. https://doi.org/10.1007/978-3-662-64553-6%5C_1
177. Podlesny, N. J., Kayem, A. V. D. M., & Meinel, C. (2022). CoK: A survey of privacy challenges in relation to data meshes. In C. Strauss et al. (Eds.), *Database and expert systems applications - 33rd international conference, DEXA 2022, Vienna, Austria, August 22-24, 2022, proceedings, part I*. Lecture Notes in Computer Science (Vol. 13426, pp. 85–102). Springer. https://doi.org/10.1007/978-3-031-12423-5_7. https://doi.org/10.1007/978-3-031-12423-5%5C_7.
178. Polonetsky, J., Tene, O., & Finch, K. (2016). Shades of gray: Seeing the full spectrum of practical data de-intentification. *Santa Clara Law Review, 56*, 593.
179. Prasser, F., et al. (2021). *ARX: Data anonymisation tool*. https://arx.deidentifier.org (visited on Oct 04, 2021).
180. Qian, J., et al. (2016). De-anonymizing social networks and inferring private attributes using knowledge graphs. In *IEEE INFOCOM 2016 - The 35th annual IEEE international conference on computer communications* (pp. 1–9). https://doi.org/10.1109/INFOCOM.2016.7524578.
181. Rolnick, D., et al. (2017). *Deep learning is robust to massive label noise*. https://doi.org/10.48550/ARXIV.1705.10694. https://arxiv.org/abs/1705.10694.
182. Romm, T. (2019). *U.S. government issues stunning rebuke, historic $ billion fine against Facebook for repeated privacy violations*. https://www.washingtonpost.com/technology/2019/07/24/us-government-issues-stunning-rebuke-historic-billion-fineagainst-facebook-repeated-privacy-violations/.
183. Rowe, F. (2020). Contact tracing apps and values dilemmas: A privacy para- dox in a neo-liberal world. *International Journal of Information Management, 55*, 102–178.
184. Rubinstein, I. S., & Hartzog, W. (2016). Anonymization and risk. In:*Washington Law Review, 91*, 703.
185. Sakpere, A. B. (2018). *Appendix A: Section A2 - CryHelp user interface questionaire* (pp. 127–131). http://hdl.handle.net/11427/28157.
186. Sakpere, A. B. (2018). *Chapter 5: User privacy preferences* (pp. 77–96). http://hdl.handle.net/11427/28157.
187. Sakpere, A. B. (2018). *Chapter 6: Implementation and experimental results* (pp. 116–119). http://hdl.handle.net/11427/28157.

188. Sakpere, A. B., & Kayem, A. V. D. M. (2014). A state-of-the-art review of data stream anonymization schemes. In *Information security in diverse computing environments* (pp. 24–50). IGI Global.
189. Sakpere, A. B., & Kayem, A. V. D. M. (2015). Supporting streaming data anonymization with expressions of user privacy preferences. In Olivier Camp et al. (Eds.), *Information systems security and privacy* (pp. 122–136). Springer International Publishing. ISBN: 978-3-319-27668-7.
190. Sakpere, A. B., & Kayem, A. V. D. M. (2017). On anonymizing streaming crime data: A solution approach for resource constrained environments. In Marit Hansen et al. (Eds.), *Privacy and identity management. The smart revolution- 12th IFIP WG 9.2, 9.5, 9.6/11.7, 11.6/SIG 9.2.2 international summer school, Ispra, Italy, September 4-8, 2017, revised selected papers*. IFIP advances in information and communication technology (Vol. 526, pp. 170–186). Springer. https://doi.org/10.1007/978-3-319-92925-5_11. https://doi.org/10.1007/978-3-319-92925-5%5C_11.
191. Sakpere, A. B., Kayem, A. V. D. M., & Ndlovu, T. (2015). A usable and secure crime reporting system for technology resource constrained context. In *29th IEEE international conference on advanced information networking and applications workshops, AINA 2015 workshops, Gwangju, South Korea, March 24–27, 2015* (pp. 424–429).
192. Samarati, P. (2001). Protecting respondents identities in microdata release. *IEEE Transactions on Knowledge and Data Engineering, 13*(6), 1010–1027.
193. Samarati, P., & Sweeney, L. (1998). *Protecting privacy when disclosing information: K-anonymity and its enforcement through generalization and suppression*. Technical Report, SRI International.
194. Sanchez, D., & Viejo, A. (2015). Privacy risk assessment of textual publications in social networks. In *Proceedings of the international conference on agents and artificial intelligence - Volume 1*. ICAART 2015. SCITEPRESS - Science and Technology Publications, Lda, 2015 (pp. 236–241). ISBN: 9789897580734. https://doi.org/10.5220/0005281202360241.
195. Sanchez, D., Martinez, S., & Domingo-Ferrer, J. (2016). Comment on "Unique in the shopping mall: On the Re-identifiability of credit card metadata". *Science (New York, N.Y.), 351*(6279), 1274. https://doi.org/10.1126/science.aad9295. ISSN: 0036-8075.
196. Sebastiani, F. (2002). Machine learning in automated text categorization. *ACM Computing Surveys, 34*(1), 1–47. https://doi.org/10.1145/505282.505283. ISSN: 0360-0300.
197. Shahriar, M. H., Kamal, A. H. , & Kayem, A. V. D. M. (2024). Discovering personally identifiable information in textual data - a case study with automated concatenation of embeddings. In L. Barolli (Ed.), *Advanced information networking and applications - Proceedings of the 38th international conference on advanced information networking and applications (AINA- 2024), Kitakyushu, Japan, 17–19 April, 2024, Volume 4*. Lecture notes on data engineering and communications technologies (Vol. 202, pp. 145–158). Springer. https://doi.org/10.1007/978-3-031-57916-5_13. https://doi.org/10.1007/978-3-031-57916-5%5C_13.
198. Shahriar, M. H., et al. (2024). Identifying personal identifiable information (PII) in unstructured text: A comparative study on transformers. In C. Strauss et al. (Eds.), *Database and expert systems applications - 35th international conference, DEXA 2024, Naples, Italy, August 26–28, 2024, proceedings, part II*. Lecture Notes in Computer Science. Springer (To Appear).
199. Shehab, M., et al. (2010). Learning based access control in online social networks. In *Proceedings of the 19th international conference on world wide web (WWW '10) Raleigh, North Carolina, USA* (pp. 1179–1180). Association for Computing Machinery. ISBN: 9781605587998. https://doi.org/10.1145/1772690.1772863.
200. da Silva, C. J. A. P. (2020). *Detecting and protecting personally identifiable information through machine learning techniques*. https://hdl.handle.net/10216/129033.
201. Sinha, A., Li, Y., & Bauer, L. (2013). What you want is not what you get: Predicting sharing policies for text-based content on facebook. In *Proceedings of the 2013 ACM workshop on artificial intelligence and security (AISec '13) Berlin, Germany* (pp. 13–24). Association for Computing Machinery. ISBN: 9781450324885. https://doi.org/10.1145/2517312.2517317.

202. Skrondal, A., & Rabe-Hesketh, S. (2003). Multilevel logistic regression for polytomous data and rankings. *Psychometrika, 68*, 267–287. https://doi.org/10.1007/BF02294801. ISSN: 1931-0145.
203. Sly, L. (2018). US soldiers are revealing sensitive and dangerous information by jogging. *The Washington Post, 29*.
204. Srinivasa-Desikan, B. (2018). *Natural language processing and computational lingustics: A practical guide to text analysis with Python, Gensim, spaCy, and Keras*.
205. Stutzman, F., Gross, R., & Acquisti, R. (2013). Silent listeners: The evolution of privacy and disclosure on facebook. *Journal of Privacy and Confidentiality, 2*, 7–41.
206. Su, J., et al. (2017). De-anonymizing web browsing data with social networks. In *Proceedings of the 26th international conference on world wide web, WWW'17* (pp. 1261–1269).
207. Sutanto, T., & Nayak, R. (2015). Semi-supervised document clustering via loci. In *International conference on web information systems engineering* (pp. 208–215). Springer.
208. Sweeney, L. (1996). Replacing personally-identifying information in medical records, the Scrub system. In *Proceedings: A conference of the American Medical Informatics Association. AMIA fall symposium* (pp. 333–337). ISSN: 1091-8280. https://europepmc.org/articles/PMC2233179.
209. Sweeney, L. (1997). Datafly: A system for providing anonymity in medical data. In *Proceedings of the IFIP TC11 WG11.3 eleventh international conference on database security XI: Status and prospects* (pp. 356–381). Chapman & Hall, Ltd. ISBN: 0-412-82090-0. http://dl.acm.org/citation.cfm?id$=$646115.679937.
210. Sweeney, L. (2000). Uniqueness of simple demographics in the US population. *LIDAP-WP4, 2000*.
211. Sweeney, L. (2000). Simple demographics often identify people uniquely. *Health (San Francisco), 671*(2000), 1–34.
212. Sweeney, L. (2002). *k*-anonymity: A model for protecting privacy. *International Journal of Uncertainty, Fuzziness and Knowledge-Based Systems, 10*(5), 557–570. ISSN: 0218-4885.
213. Sweeney, L. (2002). Achieving K-anonymity privacy protection using generalization and suppression. *International Journal of Uncertainty, Fuzziness and Knowledge-Based Systems, 10*(05), 571–588.
214. Sweeney, L., Abu, A., & Winn, J. (2013). Identifying participants in the personal genome project by name (a re-identification experiment). Preprint, arXiv:1304.7605 .
215. Terrovitis, M., Mamoulis, N., & Kalnis, P. (2008). Privacy-preserving anonymization of set-valued data. *Proceedings of the VLDB Endowment, 1*(1), 115–125.
216. Tobin, P., et al. (2017). UK financial institutions stand to lose billions in GDPR fines: How can they mitigate this? In *British Accounting And Finance Association Scottish Area Group Annual Conference, BAFA, Ed., Aberdeen* (p. 6).
217. Toch, E., Sadeh, N. M., & Hong, J. (2010). Generating default privacy policies for online social networks. In *CHI '10 extended abstracts on human factors in computing systems (CHI EA '10) Atlanta, Georgia, USA* (pp. 4243–4248). Association for Computing Machinery. ISBN: 9781605589305. https://doi.org/10.1145/1753846.1754133.
218. Tunguz, B. (1992). *Big five personality test dataset*. https://www.kaggle.com/tunguz/big-five-personality-test/version/1.
219. United Kingdom Data Service. *Text Anonymization Helper Tool*. https://ukdataservice.ac.uk/learning-hub/research-data-management/anonymisation/anonymisation-step-by-step/
220. Vaidya, J., & Clifton, C. (2003). Privacy-preserving K-means clustering over vertically partitioned data. In *Proceedings of the ninth ACM SIGKDD international conference on Knowledge discovery and data mining* (pp. 206–215). ACM.
221. Vaidya, J., Kantarcioglu, M., & Clifton, C. (2008). Privacy-preserving naive bayes classification. *The VLDB Journal—The International Journal on Very Large Data Bases, 17*(4), 879–898.
222. Verykios, V. S., et al. (2004). State-of-the-art in privacy preserving data mining. In *SIGMOD Record, 33*(1), 50–57. https://doi.org/10.1145/974121.974131. ISSN: 0163-5808.

223. Vessenes, P. J., & Seidensticker, R. B., III. (2016). *System and method for analyzing transactions in a distributed ledger*. Google Patents, 9298806. https://www.google.com/patents/US9298806.
224. Vico, H., & Calegari, D. (2015). Software architecture for document anonymization. *Electronic Notes in Theoretical Computer Science, 314*. https://doi.org/10.1016/j.entcs.2015.05.006.
225. Vuleta, B. (2021). *How much data is created every day? +27 Staggering Stats*. https://seedscientific.com/how-much-data-is-created-every-day/ (visited on Oct 10, 2022).
226. Watson, J., Lipford, H. R., & Besmer, A. (2015). Mapping user preference to privacy default settings. *ACM Transactions on Computer-Human Interaction, 22*(6), 1–20. https://doi.org/10.1145/2811257.
227. Wimmer, H., & Powell, L. (2014). A comparison of the effects of K-anonymity on machine learning algorithms. In *Proceedings of the Conference for Information Systems Applied Research ISSN* (Vol. 2167. p. 1508).
228. Witten, I. H., et al. (2017). Data mining: Practical machine learning tools and techniques. *Trees and Rules* (4th ed., Chap. 6, pp. 209– 242). Elsevier.
229. Wondracek, G., et al. (2010). A practical attack to de anonymize social network users. In *2010 IEEE symposium on security and privacy* (pp. 223–238). IEEE.
230. Wong, S. K. M., Ziarko, W., & Wong, P. C. N. (1985). Generalized vector spaces model in information retrieval. In *Proceedings of the 8th annual international ACM SIGIR conference on Research and development in information retrieval (SIGIR '85)* (pp. 18–25). Association for Computing Machinery.
231. Wong, R. C.-W., et al. (2006). (α, k)-anonymity: An enhanced k-anonymity model for privacy preserving data publishing. In *Proceed ings of the 12th ACM SIGKDD international conference on knowledge discovery and data mining (KDD '06) Philadelphia, PA, USA* (pp. 754–759). Association for Computing Machinery. ISBN: 1595933395. https://doi.org/10.1145/1150402.1150499.
232. Wong, R. C.-W., et al. (2007). Minimality attack in privacy preserving data publishing. In *Proceedings of the 33rd international conference on very large data bases (VLDB '07) Vienna, Austria* (pp. 543–554). VLDB Endowment. ISBN: 9781595936493.
233. Wong, R. C.-W., et al. (2009). Anonymization-based attacks in privacy-preserving data publishing. *ACM Transactions on Database Systems, 34*(2). ISSN: 0362-5915. https://doi.org/10.1145/1538909.1538910.
234. Wong, W. K., Mamoulis, N., & Cheung, D. W. L. (2010). Non- homogeneous generalization in privacy preserving data publishing. In *Proceedings of the 2010 ACM SIGMOD international conference on management of data* (pp. 747–758). ACM.
235. Wong, R. C.-W., et al. (2011). Can the utility of anonymized data be used for privacy breaches? *ACM Transactions on Knowledge Discovery from Data, 5*(3). ISSN: 1556-4681. https://doi.org/10.1145/1993077.1993080.
236. Xiao, X., & Tao, Y. (2006). Personalized privacy preservation. In *Proceedings of the 2006 ACM SIGMOD international conference on management of data (SIGMOD '06) Chicago, IL, USA* (pp. 229–240). Association for Computing Machinery. ISBN: 1595934340. https://doi.org/10.1145/1142473.1142500.
237. Xiao, X., Ke, Y., & Tao, Y. (2010). The hardness and approximation algorithms for L-diversity. In *Proceedings of the 13th international conference on extending database technology (EDBT '10). Lausanne, Switzerland* (pp. 135–146). ACM. ISBN: 978-1-60558-945-9.
238. Xiao, Q., Reiter, M. K., & Zhang, Y. (2015). Mitigating storage side channels using statistical privacy mechanisms. In *Proceedings of the 22nd ACM SIGSAC conference on computer and communications security (CCS '15) Denver, Colorado, USA* (pp. 1582–1594). ACM. ISBN: 978-1-4503-3832-5.
239. Yang, X., & Hayashi, T. (2021). Exploring the effects of internet memes in social media marketing through A/B testing. In *2021 IEEE 23rd conference on business informatics (CBI)* (Vol. 02, pp. 97–106). https://doi.org/10.1109/CBI52690.2021.10060.

240. Yang, X., & Li, C. (2004). Secure XML publishing without information leakage in the presence of data inference. In *Proceedings of the thirtieth international conference on very large data bases* (Vol. 30, pp. 96–107).
241. Yuan, M., Chen, L., & Yu, P. S. (2010). Personalized privacy protection in social networks. In *Proceeding of the VLDB Endowment, 4*(2), pp. 141–150. https://doi.org/10.14778/1921071.1921080. ISSN: 2150-8097.
242. Zakerzadeh, H., & Osborn, S. L. (2013). Delay-sensitive approaches for anonymizing numerical streaming data. In *International Journal of Information Security, 12*, 423–437.
243. Zhang, B., et al. (2015). Feature selection for classification under anonymity constraint. Preprint, arXiv:1512.07158.
244. Zimmermann, V., & Renaud, K. (2021). The nudge puzzle: Matching nudge interventions to cybersecurity decisions. In *ACM Transactions on Computer-Human Interaction, 28*(1). ISSN: 1073-0516. https://doi.org/10.1145/3429888.

Index

A. V. D. M. Kayem, *Enabling Privacy Preserving Data Analytics*, Advances in Information Security 92, https://doi.org/10.1007/978-3-031-93906-8

9783031939051